The Direct Method in Soliton Theory

CAMBRIDGE TRACTS IN MATHEMATICS

155 The Direct Method in Soliton Theory

The Direct Method in Soliton Theory

RYOGO HIROTA

Translated from Japanese and edited by Atsushi Nagai,
Jon Nimmo and Claire Gilson

CAMBRIDGE
UNIVERSITY PRESS

CAMBRIDGE
UNIVERSITY PRESS

University Printing House, Cambridge CB2 8BS, United Kingdom

Published in the United States of America by Cambridge University Press, New York

Cambridge University Press is part of the University of Cambridge.

It furthers the University's mission by disseminating knowledge in the pursuit of education, learning and research at the highest international levels of excellence.

www.cambridge.org
Information on this title: www.cambridge.org/9780521836609

First published in Japanese as *Chokusetsuhō ni yoru soriton no sūri* by Iwanami Shoten in 1992

First published in English 2004

A catalogue record for this publication is available from the British Library

Library of Congress Cataloguing in Publication data
Hirota, Ryogo, 1932–
[Chokusetsuho ni yoru soriton no suri. English]
The direct method in soliton theory/Ryogo Hirota.
p. cm. – (Cambridge tracts in mathematics; 155)
Includes bibliographical references and index.
ISBN 0 521 83660 3
1. Solitons. I. Title. II. Series.
QC174.26.W28H4713 2004
530.12'4 – dc22 2003069569

ISBN 978-0-521-83660-9 Hardback

Contents

Foreword

The second half of the twentieth century saw a resurgence in the study of classical physics. Scientists began paying particular attention to the effects caused by the nonlinearity in dynamical equations. This nonlinearity was found to have two interesting manifestations of opposite nature: *chaos*, that is the apparent randomness in the behaviour of perfectly deterministic systems, and *solitons*, that is localized, stable moving objects that scattered elastically. Both of these topics have now been developed into paradigms, with solid mathematical background and with a wide range of physical observations and concrete applications.

This book is concerned with a particular method used in the study of solitons. There are many ways of studying the integrable nonlinear evolution equations that have soliton solutions, each method having its own assumptions and areas of applicability. For example, the inverse scattering transform (IST) can be used to solve initial value problems, but it uses powerful analytical methods and therefore makes strong assumptions about the nonlinear equations. On the other hand, one can find a travelling wave solution to almost all equations by a simple substitution which reduces the equation to an ordinary differential equation. Between these two extremes lies Hirota's direct method. Although the transformation was, at its heart, inspired by the IST, Hirota's method does not need the same mathematical assumption and, as a consequence, the method is applicable to a wider class of equations than the IST. At the same time, because it does not use such sophisticated techniques, it usually produces a smaller class of solutions, the multi-soliton solutions. In many problems the key to further developments is a detailed understanding of soliton scattering, and in such cases Hirota's bilinear method is the optimal tool.

Over the years, many textbooks have been written on various aspects of solitons. Although some of them have briefly mentioned Hirota's bilinear method, there has not been any introductory English language book devoted to it. When some of the western practitioners of Hirota's method found out about prof. Hirota's book, they felt that it should be translated into English for use as an introduction to the method. Early in 1997, Prof. J. Satsuma recruited his students A. Nagai for this translation project. With help from his colleagues, S. Tsujimoto and R. Willox, a first version was made; this was further improved in a collaboration between A. Nagai, J. J. C. Nimmo and C. R. Gilson, leading to the final version that is presented here.

In this book, Prof. Hirota explains his 'direct' or 'bilinear' method. There is an interesting introduction, from which we can see the motivation and chain of thought that led Prof. Hirota to invent his method. The rest of the book is devoted to a detailed discussion of various applications of the method. Little has been changed in the translation from the original book; one or two arguments have been expanded, some errors have been corrected and some notation was changed to improve consistency. All such changes have been made with the approval of Prof. Hirota.

Jarmo Hietarinta
Turku, Finland

Jon Nimmo
Glasgow, Scotland

Preface

A soliton is a particular type of solitary wave, which is not destroyed when it collides with another wave of the same kind. Such behaviour is suggested by numerical simulation, but is it really possible that the soliton completely recovers its original shape after a collision? In detailed analysis of the results of such numerical simulations, some ripples can be observed after a collision, and it therefore seems that the original shape is not completely recovered. Therefore, in order to clarify whether or not solitons are destroyed through their collisions, it is necessary to find exact solutions of soliton equations.

Generally, it is a very hard task to find exact solutions of nonlinear partial differential equations, including soliton equations. Moreover, even if one manages to find a method for solving one nonlinear equation, in general such a method will not be applicable to other equations. Does there exist any successful and universal tool enabling one to solve many types of nonlinear equations which does not require a deep understanding of mathematics? For this purpose, a direct method has been investigated.

In Chapter 1, we discuss in an intuitive way the conditions under which a solitary wave is formed and we show that a nonlinear solitary wave cannot be made by the superposition of linear waves. From this observation, we obtain a method for finding solutions of a nonlinear wave equation by reductive perturbations and derive the fundamental idea for the direct method. We first introduce new dependent variables F and G to express the solution of the equation as the ratio G/F and then solve the (now bilinear) equations for F and G. As part of this, a new binary operator, called the D-operator, is derived. General formulae, through which nonlinear partial differential equations are transformed into bilinear (or, in general, homogeneous) forms, are presented. By virtue of special properties of the D-operator, solving these bilinear forms by ordinary reductive perturbation methods leads to perturbation expansions that may sometimes be truncated as finite sums. Such a truncation yields an exact solution for the equation. As an example, we find an exact solution for one of the most famous soliton equations, the KdV equation, and prove that its solitons are preserved after interaction.

In Chapter 2, we introduce the mathematical tools – in particular the theory of determinants and pfaffians – to be used in Chapter 3. These techniques will be explained thoroughly by means of several examples so that readers with only elementary knowledge can understand them. Consequently, this chapter covers one-quarter of the book.

In Chapter 3, we discuss the structure of soliton equations from the viewpoint of the direct method presented in this book. Many kinds of soliton equations have been discovered up to now and it would require several pages to write them all down. Now the question arises: what is the fundamental structure common to all soliton equations? The answer is provided in this chapter; soliton equations (or bilinear forms) are nothing but 'pfaffian identities'. From this viewpoint, we show how fundamental soliton equations, such as the KP, BKP, coupled KP, Toda lattice and Toda molecule equations, resolve themselves into pfaffian identities.

Pfaffians, which may be an unfamiliar word, are closely related to determinants. They are usually defined by the property that the square of a pfaffian is the determinant of an antisymmetric matrix. This property often gives rise to the misunderstanding that a pfaffian is merely a special case of a determinant. In fact, it is more natural to regard a pfaffian as a *generalization* of a determinant. For example, Plücker relations and Jacobi identities, which are identities for determinants, also hold for pfaffians. As a matter of fact, they can be extended and unified as pfaffian identities.

By means of the Maya diagrams designed by Professor Mikio Sato, a pfaffian identity can be illustrated by the formula

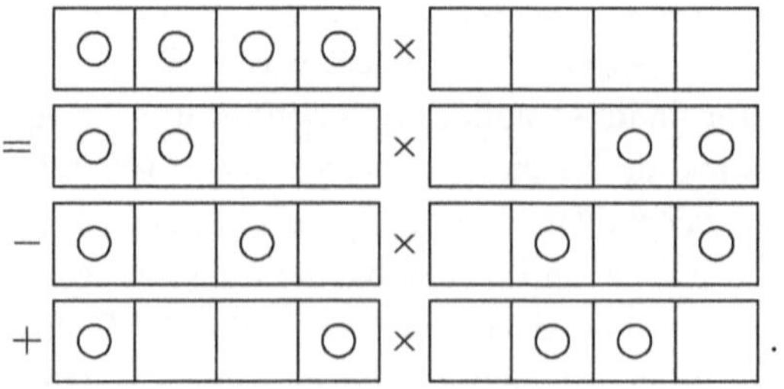

It is quite a surprise that soliton equations reduce to such simple diagrams!

In Chapter 4, we discuss Bäcklund transformations, which have made important contributions in the development of soliton theory. Bäcklund transformations in bilinear form generate (i) Lax pairs used in the inverse scattering method, (ii) new soliton equations, and (iii) Miura transformations. A Bäcklund transformation in bilinear forms corresponds to an 'exchange formula' for the D-operator. First, we find a Bäcklund transformation for the KdV equation by using such an exchange formula. Next, we illustrate some applications of this Bäcklund transformation for the KdV equation. Finally, we clarify the structure of Bäcklund transformations for other soliton equations such as the KP, BKP and Toda equations, and we also show that all these Bäcklund transformations reduce to pfaffian identities.

Since most of this book is devoted to an explanation of the fundamental facts concerning the direct method, space does not permit us to mention its applications in many other fields. It is particularly disappointing that we could not touch upon the group-theoretical aspects of bilinear forms developed by the Sato school (Professors Mikio Sato, Yasuko Sato, Masaki Kashiwara, Tetsuji Miwa, Michio Jimbo and Etsuro Date). With regard to inverse scattering methods, we have completely omitted them because many books have already been written on this subject. The aim of this book is to inform the readers as briefly as possible about the beauty and conciseness of the mathematical rules underlying soliton equations.

The author is greatly indebted to members of Professor Mikio Sato's school in Kyoto University and those of Professor Junkichi Satsuma's laboratory in the University of Tokyo, for their own developments of direct methods. He also thanks Dr Hideyuki Kidachi, whose notes on the author's lectures (Department of Physics, Faculty of Science, Kyoto University, 1–3 February 1979) were very useful in writing Chapter 1. Last but not least, the author is grateful to Mr Satoshi Tsujimoto and Mr Tatsuya Imai for their help drawing figures and proofreading.

1

Bilinearization of soliton equations

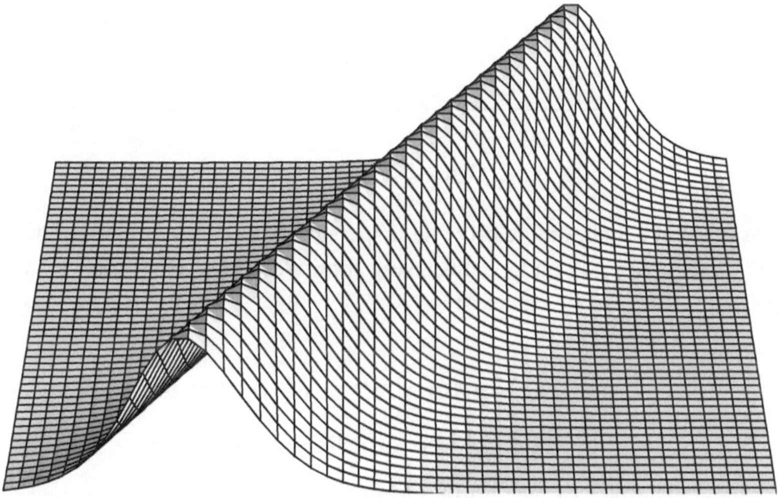

One-soliton solution.

1.1 Solitary waves and solitons

The word 'wave' normally makes us think of a wave train as shown in Figure 1.1. However, when surfing off a gently sloping beach, we often make use of a *solitary wave* (see Figure 1.2). A *soliton* is a type of solitary wave which maintains its identity after it collides with another wave of the same kind. Let us first study wave equations which describe solitary waves.

A wave equation having soliton solutions has both nonlinearity and dispersion. Before studying how to solve such a wave equation, we will investigate

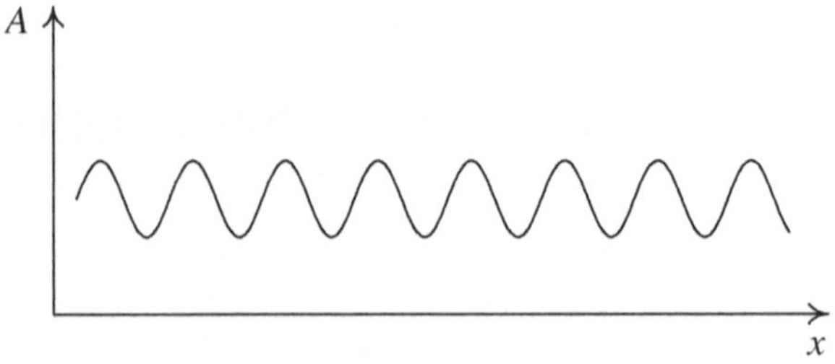

Figure 1.1. A wave train. Amplitude A, position x.

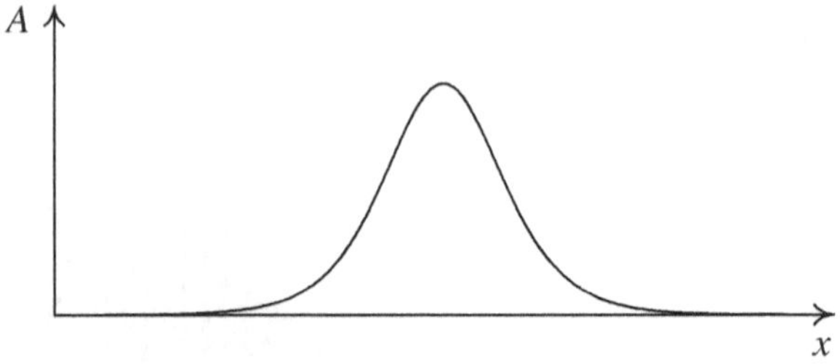

Figure 1.2. A solitary wave.

the influence that nonlinearity and dispersion have on the behaviour of a wave. We will also try to understand, using intuitive arguments, under what conditions a solitary wave can exist.

1.2 Nonlinearity and dispersion

1.2.1 Linear nondispersive waves

Typical examples of the simplest kind of waves are sound waves and electromagnetic waves. They are governed by

$$\left(\frac{\partial^2}{\partial t^2} - v_0^2 \frac{\partial^2}{\partial x^2}\right) f(x,t) = 0, \tag{1.1}$$

where v_0 is a constant representing the wave speed. Since this equation can be formally decomposed as

$$\left(\frac{\partial}{\partial t} - v_0 \frac{\partial}{\partial x}\right)\left(\frac{\partial}{\partial t} + v_0 \frac{\partial}{\partial x}\right) f(x,t) = 0,$$

let us consider the simpler form,

$$\left(\frac{\partial}{\partial t} + v_0 \frac{\partial}{\partial x}\right) f(x,t) = 0. \tag{1.2}$$

A solution of this equation also satisfies (1.1). While (1.1) gives travelling wave solutions moving to the left and to the right, (1.2) gives only the right-moving ones,

$$f(x,t) = f(x - v_0 t).$$

Assuming that this wave is periodic, the most fundamental solution is the plane wave,

$$f(x,t) = \exp[\mathrm{i}(\omega t - kx)].$$

The relationship between the angular frequency ω and the wave number k is given by $\omega = v_0 k$, where the constant v_0 is the phase velocity of the wave. This is called the *dispersion relation* and, in this case, it is linear.

A wave governed by a linear dispersion relation is called a *nondispersive wave*. A feature of such a wave is that an initial profile taking the form of a pulse, which is made up of a superposition of plane waves with different wave numbers k, does not change its shape. This is because each of the superposed plane waves travels with the same speed. Waves with unchanging shape play a very important role in applications as a means of communication. A soliton, even though it is not a nondispersive wave, possesses the above property of unchanging shape and, because of this, it should have practical applications.

Next, we will investigate a particular linear dispersive wave equation.

1.2.2 Linear dispersive waves

We consider, as the simplest example, the wave equation

$$\left(\frac{\partial}{\partial t} + v_0 \frac{\partial}{\partial x} + \delta \frac{\partial^3}{\partial x^3}\right) f(x,t) = 0. \tag{1.3}$$

If we suppose that it has a plane wave solution

$$f(x,t) \propto \exp[\mathrm{i}(\omega t - kx)],$$

then the dispersion relation is given by

$$\omega = v_0 k - \delta k^3,$$

which is nonlinear with respect to k. Hence, the phase velocity is different from that in Section 1.2.1 and is given by

$$\frac{\omega}{k} = v_0 - \delta k^2,$$

which depends on a wave number k. On the other hand, its group velocity is given by

$$\frac{\partial \omega}{\partial k} = v_0 - 3\delta k^2.$$

We remark that, if $\delta > 0$, both velocities are less than v_0. Since the velocity of each of the plane waves which make up an initial wave vary with k, the wave spreads out as it travels. This shows that linear dispersive waves do not preserve their original shape.

The two examples we have discussed so far are both linear differential equations. Next we consider the influence of nonlinearity.

1.2.3 Nonlinear nondispersive waves

We consider, as the simplest example, the nondispersive wave equation,

$$\left(\frac{\partial}{\partial t} + v(f)\frac{\partial}{\partial x}\right) f(x,t) = 0, \tag{1.4}$$

where $v(f) = v_0 + \alpha f^m$. This equation is a nonlinear wave equation in which the speed $v(f)$ depends on the amplitude f.

Equation (1.4) has the formal solution

$$f(x,t) = f(x - v(f)t),$$

and if $v = v(f)$ is an increasing function in f, this formula tells us that a wave travels faster as its amplitude increases. Therefore, as one can see from Figure 1.3, the wave steepens and then breaks. Physically speaking, however, before the wave breaks, its gradient $|\partial f/\partial x| \gg 1$. When this happens, (1.4) becomes meaningless and must be replaced by the differential equation (1.5) presented in Section 1.2.4.

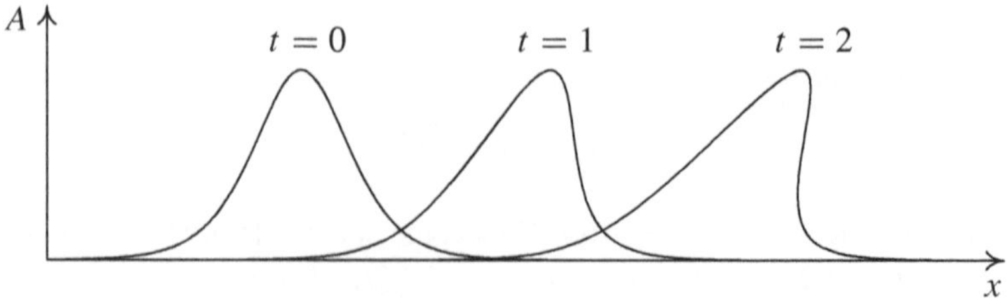

Figure 1.3. Steepening of a solitary wave. A wave which is symmetrical at $t = 0$ steepens and breaks because of the dependence of the wave speed on its amplitude.

1.2.4 Nonlinear dispersive waves

The equation

$$\left(\frac{\partial}{\partial t} + v_0\frac{\partial}{\partial x} + \alpha f^m\frac{\partial}{\partial x} + \delta\frac{\partial^3}{\partial x^3}\right) f(x,t) = 0 \tag{1.5}$$

possesses a pulse-like wave solution which travels with unchanging shape; a solitary wave solution. Before finding this solution by mathematical means, let us look into the physical reasons for the existence of such a solution. We have seen in Sections 1.2.2 and 1.2.3 that neither a linear dispersive solitary wave nor a nonlinear nondispersive solitary wave can exist. Why then can a solitary wave solution exist for a wave equation which has both nonlinearity and dispersion?

Assuming that the solitary wave shown in Figure 1.4 exists, let us investigate whether this pulse-like wave can travel with unchanging shape. To this end it is necessary, at least, that the velocities at the top and base of the wave have the same value v. In order to investigate further, we introduce a new space coordinate $\eta = px - \Omega t$, where $v = \Omega/p$ and p is a free parameter. The parameter p is such that, as it increases, the pulse becomes sharper. It is convenient to introduce η to describe a wave which travels at a constant speed v. For $t = 0$, η is proportional to x ($\eta = px$), and, for $t \neq 0$, $\eta = p(x - vt)$, being proportional to $x - vt$, travels at a speed v.

If the maximum of the wave amplitude f is A, occurring at $\eta = 0$, then in a neighbourhood of this point we have

$$f \sim A(1 - \text{constant} \times \eta^2),$$

because here the height f can be approximated by a quadratic expression in η. From this equation, we have $\partial^3 f/\partial x^3 \sim 0$ and therefore, in the neighbourhood

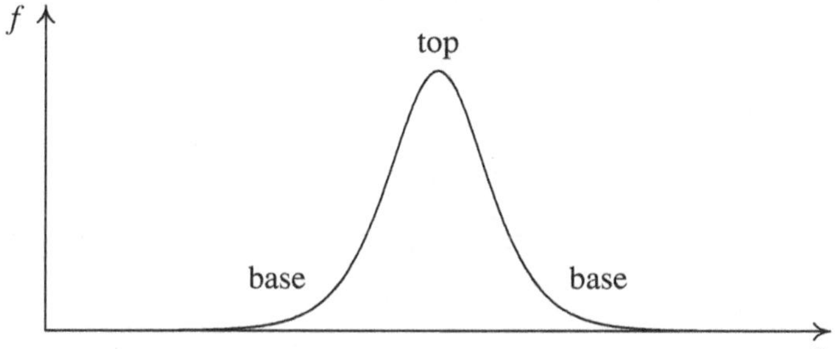

Figure 1.4. Splitting a solitary wave into its top and base.

of the top of the wave, f satisfies the differential equation

$$\left(\frac{\partial}{\partial t} + [v_0 + \alpha f(x,t)^m]\frac{\partial}{\partial x}\right) f(x,t) \sim 0.$$

This equation is the same as (1.4) and so, as described in Section 1.2.3, the speed at the top of the wave, at which the amplitude is A, is given by

$$v(f) = v_0 + \alpha A^m. \tag{1.6}$$

From this it is clear that $v(f)$ is larger than v_0 if $\alpha > 0$.

On the other hand, at the base of the wave, we can neglect the nonlinear term because f is very small, and so f satisfies the linear differential equation

$$\left(\frac{\partial}{\partial t} + v_0\frac{\partial}{\partial x} + \delta\frac{\partial^3}{\partial x^3}\right) f(x,t) \sim 0. \tag{1.7}$$

As seen in Section 1.2.2, the group and phase velocities at the base, $v_{\rm gr}$ and $v_{\rm pf}$, respectively, are given by

$$\begin{aligned} v_{\rm gr} &= \frac{\partial \omega}{\partial k} = v_0 - 3\delta k^2, \\ v_{\rm pf} &= \frac{\omega}{k} = v_0 - \delta k^2. \end{aligned} \tag{1.8}$$

From this we see that both the group and phase velocities are smaller than v_0 if $\delta > 0$, and so the speed at the top of a solitary wave is larger than that at the base. This indicates that the original shape of the wave is not preserved, and a solitary wave cannot exist. This disagrees with the experimental observation. What is wrong with the above argument?

In fact, our calculation of the velocity at the base of the wave is incorrect. Since the amplitude at the base is small, the differential equation is certainly linear. The error arose because we approximated the solution as a plane wave and considered the wave velocity to be the linear (phase or group) velocity, according to the common understanding of linear waves. The base of the wave is not made up of a superposition of linear plane waves

$$f \sim \exp[\pm {\rm i}(kx - \omega t)]$$

or

$$f \sim \sin(kx - \omega t),$$

but, in fact, is expressed in terms of exponentially decaying waves

$$f \sim \exp[\pm(px - \Omega t)]. \tag{1.9}$$

Since these expressions for f tend to infinity as $x \to \infty$ or $x \to -\infty$, and so do not satisfy a physical boundary condition, solutions of this form are

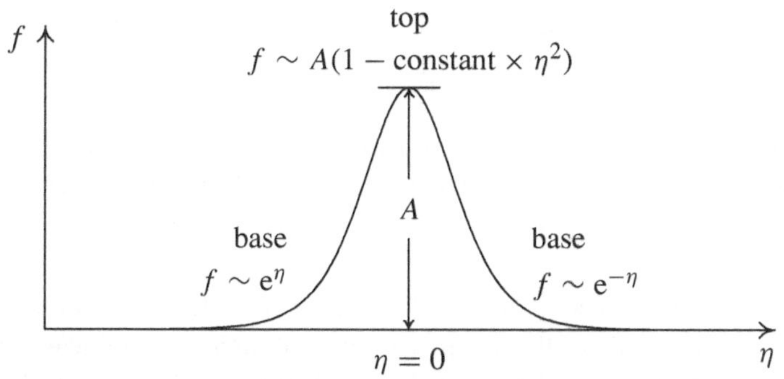

Figure 1.5. Approximation of a solitary wave at its top and base.

normally discarded as physically meaningless in the theory of linear waves. In the theory of nonlinear waves, however, we can construct a global solution by connecting local solutions, as illustrated in Figure 1.5.

The base of the wave is now expressed in terms of exponentially decaying solutions $f \sim \exp(\pm\eta)$ and, from (1.7), we obtain the relationship, called the *nonlinear dispersion relation*,

$$\Omega = v_0 p + \delta p^3.$$

The wave velocity v is then given by

$$v = \frac{\Omega}{p} = v_0 + \delta p^2, \tag{1.10}$$

which coincides with the velocity at the top (1.6) if and only if

$$\delta p^2 = \alpha A^m.$$

If p and A satisfy this equation, then the solitary wave can travel without changing its shape. Recall that p is the parameter associated with the width of the pulse; as p increases, the pulse becomes steeper and narrower. This formula (if $\delta, \alpha, m > 0$) therefore indicates that as the amplitude of the pulse increases, it becomes sharper.

The above discussion suggests an important idea for solving nonlinear wave equations. When trying to obtain a solution by a perturbation method, we cannot employ, as a first approximation, the normal plane wave solutions $f \sim \sin(kx - \omega t)$ but should instead use the exponentially decaying solutions $f \sim \exp[\pm(px - \Omega t)]$, which are rejected in linear wave theory. More precisely, we expand f into a power series in $\varepsilon \exp(\eta)$ as

$$f(x,t) \sim \varepsilon a_1 \exp(\eta) + \varepsilon^2 a_2 \exp(2\eta) + \cdots, \tag{1.11}$$

where $\eta = px - \Omega t$ and ε is a small parameter.

However, if η is sufficiently large, we have seen (see Figure 1.5) that

$$f(x,t) \sim \exp(-\eta). \tag{1.12}$$

Expanding f into a power series of $\exp(\eta)$ and finding a solution asymptotic to $\exp(-\eta)$ as $\eta \to +\infty$ corresponds to finding a Padé approximation for f,

$$f = G/F. \tag{1.13}$$

If this correspondence is correct, the fundamental idea of the direct method, referred to in the title of this book, is to find solutions of nonlinear differential equations through dependent variable transformations like $f = G/F$.

Remark

For a function f with formal power series

$$f(x) = a_0 + a_1 x + a_2 x^2 + \cdots, \tag{1.14}$$

the method of Padé approximation [1] expresses $f(x)$ as a ratio of polynomials G and F. This gives an approximate analytic continuation for $f(x)$ that can be used to obtain information for large x.

For example, the power series

$$f(x) = x - x^3 + x^5 - x^7 + \cdots \tag{1.15}$$

converges to a finite value in a region $|x| < 1$ and therefore gives properties of $f(x)$ in this region. In the region $|x| \geq 1$, however, this is a divergent series and so it does not make sense to use it, for example, to find the value at $x = 2$. If we express $f(x)$ as the rational function

$$f(x) = \frac{x}{1+x^2},$$

then $f(x) \sim x^{-1}$ for $|x| \gg 1$. In particular, the substitution of $x = \exp(\eta)$ yields

$$\begin{aligned} f(x) &= \exp(\eta) - \exp(3\eta) + \exp(5\eta) - \cdots \\ &= \frac{\exp(\eta)}{1+\exp(2\eta)} \\ &\sim \exp(-\eta) \quad (\eta \gg 1) \end{aligned} \tag{1.16}$$

and gives the correct behaviour of $f(x)$, even though η is large. □

Here we have used an intuitive argument to find general properties of the solitary wave solution without any knowledge of the precise form of the nonlinear term. In Section 1.3 we will investigate to what extent this solution coincides with the exact solution.

1.3 Solutions of nonlinear differential equations

In Section 1.2 we discussed solutions of the nonlinear differential equation

$$\left(\frac{\partial}{\partial t'} + v_0\frac{\partial}{\partial x'} + \alpha f^m \frac{\partial}{\partial x'} + \delta\frac{\partial^3}{\partial x'^3}\right) f(x', t') = 0, \tag{1.17}$$

using an intuitive argument (we use variables x', t' for later convenience). Let us here investigate the properties of such solutions by mathematical means.

First, we consider the independent variable transformation

$$\begin{aligned} x &= x' - v_0 t', \\ t &= t', \end{aligned} \tag{1.18}$$

describing a frame moving at velocity v_0. Under this transformation, partial derivatives become

$$\begin{aligned} \frac{\partial}{\partial t'} &= \frac{\partial}{\partial t} - v_0\frac{\partial}{\partial x}, \\ \frac{\partial}{\partial x'} &= \frac{\partial}{\partial x}. \end{aligned} \tag{1.19}$$

In the moving frame, the second term in (1.17) is eliminated and so we obtain

$$\left(\frac{\partial}{\partial t} + \alpha f^m \frac{\partial}{\partial x} + \delta\frac{\partial^3}{\partial x^3}\right) f(x, t) = 0. \tag{1.20}$$

Next we make use of a *similarity transformation*. Under the scaling transformation $t = \varepsilon^3\tau, x = \varepsilon\xi$, where ε is a constant, (1.20) is equivalent to

$$\left(\frac{\partial}{\partial \tau} + \varepsilon^2\alpha f^m \frac{\partial}{\partial \xi} + \delta\frac{\partial^3}{\partial \xi^3}\right) f(\varepsilon\xi, \varepsilon^3\tau) = 0, \tag{1.21}$$

and then the dependent variable transformation $f(\varepsilon\xi, \varepsilon^3\tau) = \varepsilon^{-2/m} f'(\xi, \tau)$ yields

$$\left(\frac{\partial}{\partial \tau} + \alpha (f')^m \frac{\partial}{\partial \xi} + \delta\frac{\partial^3}{\partial \xi^3}\right) f'(\xi, \tau) = 0. \tag{1.22}$$

This shows that if we replace f in (1.20) by

$$f'(\xi, \tau) = \varepsilon^{2/m} f(\varepsilon\xi, \varepsilon^3\tau), \tag{1.23}$$

then f' again satisfies the same differential equation. This is called a similarity transformation.

If we have a travelling wave solution $f = f(x - vt)$ then $\varepsilon^{2/m} f(x - \varepsilon^2 vt)$ will also be a solution. From this we see that if the amplitude of a solitary wave increases by a factor $\varepsilon^{2/m}$ then its velocity v increases by a factor of ε^2. This shows that, even in the case that an exact solution cannot be found, we can still determine some properties of these solutions by their similarity transformations. In fact, exact analytic solutions for few nonlinear differential equations are known.

The relation between the speed and amplitude of solitary waves can also be found directly from the partial differential equation without employing a similarity transformation. Let us again consider the nonlinear partial differential equation

$$\left(\frac{\partial}{\partial t} + v_0 \frac{\partial}{\partial x} + \alpha f^m \frac{\partial}{\partial x} + \delta \frac{\partial^3}{\partial x^3}\right) f(x, t) = 0. \tag{1.24}$$

We consider a solitary wave solution $f = f(x - vt)$ travelling at constant speed v. Then we have

$$\frac{\partial f}{\partial t} = -v \frac{\partial f}{\partial x}, \tag{1.25}$$

and (1.24) reduces to

$$\left((-v + v_0) \frac{\partial}{\partial x} + \alpha f^m \frac{\partial}{\partial x} + \delta \frac{\partial^3}{\partial x^3}\right) f(x - vt) = 0. \tag{1.26}$$

Integrating the above equation with respect to x and using the boundary condition for solitary waves,

$$\frac{\partial^n}{\partial x^n} f(x - vt) \to 0 \quad (x \to \pm\infty) \quad n = 0, 1, 2, \ldots, \tag{1.27}$$

we have

$$(-v + v_0) f + \frac{\alpha}{m+1} f^{m+1} + \delta \frac{\partial^2}{\partial x^2} f = 0. \tag{1.28}$$

Multiplying by f_x on both sides and integrating with respect to x again, we obtain

$$(-v + v_0) f^2 + \frac{2\alpha}{(m+1)(m+2)} f^{m+2} + \delta f_x^2 = 0. \tag{1.29}$$

At the top of the solitary wave we suppose that $f = f_{\max}$, and we have

$$f_x = 0, \tag{1.30}$$

which gives

$$(-v + v_0) f_{\max}^2 + \frac{2\alpha}{(m+1)(m+2)} f_{\max}^{m+2} = 0. \tag{1.31}$$

Hence, we may deduce that the relationship between the velocity v and height $f_{\max}$ is

$$v = v_0 + \frac{2\alpha}{(m+1)(m+2)} f_{\max}^m. \tag{1.32}$$

On the other hand, an exact solitary wave solution for (1.24) is given by

$$f = A(\cosh \eta)^{-s}, \quad \eta = px - \Omega t + \text{constant}, \tag{1.33}$$

where $\Omega = v_0 p + \delta s^2 p^3$, $v = \Omega/p = v_0 + \delta s^2 p^2$ and $s = 2/m$. As a consequence of (1.32), we have $A^m = (\delta/\alpha)(1+s)(2+s)p^2$. This relationship between the wave height A and the parameter p should be compared with the relation

$$A^m = (\delta/\alpha) p^2, \tag{1.34}$$

obtained by an intuitive argument in the previous section.

Remarks

(1) When one can find an exact solution, we might think that an intuitive discussion of the solution is unnecessary. However, exact solutions are known for only a very limited class of equations. For example, an analytic solitary wave solution for the nonlinear differential equation

$$\left(\frac{\partial}{\partial t} + \alpha f \frac{\partial}{\partial x} + \delta \frac{\partial^5}{\partial x^5} \right) f(x,t) = 0 \tag{1.35}$$

has not yet been found. In cases where an exact solution is not known, we can still obtain a rough impression of the solution by making use of an intuitive argument.

(2) The word *soliton* was first used in the paper published by Zabusky and Kruskal in 1965 [2]. They carried out a numerical experiment on the Korteweg–de Vries (KdV) equation,

$$\left(\frac{\partial}{\partial t} + \alpha v \frac{\partial}{\partial x} + \delta \frac{\partial^3}{\partial x^3} \right) v(x,t) = 0, \tag{1.36}$$

which describes shallow water waves. In doing this they discovered that solitary waves are not destroyed when they collide. They called the solitary wave solution a soliton, the suffix -on denoting a particle to indicate particle-like interaction properties.

In order to prove rigorously that solitary wave solutions for the KdV equation are not destroyed after their interaction, it is necessary to find exact solutions describing this interaction. Gardner, Greene, Kruskal and Miura discovered the inverse scattering method [3] to find soliton solutions for the KdV equation in 1967.

(3) The direct method, which is the topic of this book, is a method for finding soliton solutions directly, without employing the inverse scattering method. □

The important point that this section demonstrates is that, even though both nonlinearity and dispersion on their own destroy pulse-like waves, acting together they balance each other out and permit such waves to exist.

1.4 Linearization of nonlinear differential equations

Since the superposition principle holds in the case of linear differential equations, it is relatively easy to construct general solutions from particular solutions. On the other hand, it is very difficult to find exact solutions of nonlinear differential equations.

In the case of nonlinear differential equations, the superposition principle does not hold and there may exist solutions which could not be obtained by guesswork.

One way to solve a nonlinear differential equation is to find a transformation to a linear equation. We are, however, unable to categorize the nonlinear differential equations that can be linearized; all we can do is to give some examples.

1.4.1 The Riccati equation

The Riccati equation,

$$\frac{\mathrm{d}}{\mathrm{d}t}u(t) = a(t) + 2b(t)u(t) + u(t)^2, \tag{1.37}$$

is one of the simplest nonlinear first-order ordinary differential equations. It may be linearized by making the change of dependent variable

$$u = g/f. \tag{1.38}$$

Differentiation with respect to t gives

$$u_t = \frac{g_t f - g f_t}{f^2}, \tag{1.39}$$

where subscripts denote derivatives. Substitution into (1.37) yields

$$g_t f - g f_t = a(t) f^2 + 2b(t) f g + g^2, \tag{1.40}$$

which may be rearranged to give

$$[g_t - a(t) f - b(t) g] f - [f_t + b(t) f + g] g = 0. \tag{1.41}$$

Introducing an arbitrary function $\lambda(t)$, we can separate (1.41) into two coupled linear differential equations,

$$\begin{aligned} f_t + b(t) f + g &= \lambda(t) f, \\ g_t - a(t) f - b(t) g &= \lambda(t) g. \end{aligned} \tag{1.42}$$

This is a linearization of the Riccati equation.

The same situation, in which a nonlinear differential equation is rewritten as (coupled) linear differential equations by means of a dependent variable transformation to a rational function g/f, is also found for nonlinear partial differential equations such as the Burgers equation.

1.4.2 The Burgers equation

The Burgers equation,

$$u_t = u_{xx} + 2uu_x, \tag{1.43}$$

can be integrated with respect to x by introducing a potential function w, where $u = w_x$. It then becomes

$$w_t = w_{xx} + w_x^2 + c, \tag{1.44}$$

where c is an arbitrary constant of integration. Through a dependent variable transformation $w = \log f$, this equation gives

$$\frac{f_t}{f} = \frac{f_{xx} f - f_x^2}{f^2} + \frac{f_x^2}{f^2} + c.$$

Since the second and the third terms on the right-hand side cancel each other, multiplication through by f gives a linear equation,

$$f_t = f_{xx} + cf. \tag{1.45}$$

The dependent variable transformation used here,

$$u = (\log f)_x = f_x / f, \tag{1.46}$$

is called the *Cole–Hopf transformation* [4]. In summary, the Burgers equation $u_t = u_{xx} + 2uu_x$ is transformed into the linear partial differential equation $f_t = f_{xx} + cf$ through the Cole–Hopf transformation $u = (\log f)_x = f_x/f$.

Considering the inverse of this procedure, we can derive a nonlinear differential equation starting from a linear differential equation. For example, linear partial differential equations,

$$\frac{\partial f}{\partial t} = \frac{\partial^n f}{\partial x^n} \quad (n = 2, 3, 4, \ldots), \tag{1.47}$$

are transformed into nonlinear partial differential equations through the inverse of the Cole–Hopf transformation,

$$f = \exp\left(\int u \, \mathrm{d}x\right). \tag{1.48}$$

We obtain the Burgers equation if $n = 2$, and if $n = 3$ we have the nonlinear partial differential equation

$$u_t = u_{xxx} + 3uu_{xx} + 3u_x^2 + 3u^2 u_x. \tag{1.49}$$

An equation similar to the above,

$$u_t = u_{xxx} + \alpha uu_x + \beta u^2 u_x, \tag{1.50}$$

is a soliton equation, called the modified Korteweg–de Vries (mKdV) equation. However, these two equations have solutions with completely different structures.

Next we consider the differential-difference equation

$$\frac{\partial}{\partial t} u_n = (1 + u_n)(u_{n+1} - u_n), \tag{1.51}$$

in which a discrete variable n is employed instead of a continuous independent variable x. The above equation may also be written as

$$\frac{\partial}{\partial t} \log(1 + u_n) = (u_{n+1} - u_n). \tag{1.52}$$

Since the left-hand side is written as the t-derivative of a logarithm, we may assume that u_n on the right-hand side may be written in the same way. Through the dependent variable transformation

$$u_n = (\log f_n)_t, \tag{1.53}$$

we have, from (1.52),

$$\frac{\partial}{\partial t}\log[1+(\log f_n)_t] = \frac{\partial}{\partial t}\log\frac{f_{n+1}}{f_n}. \tag{1.54}$$

Integrating with respect to t and taking an exponential on both sides, the above equation gives the linear equation

$$\frac{\partial}{\partial t} f_n = c f_{n+1} - f_n, \tag{1.55}$$

where c is a constant of integration.

Remark

The differential-difference equation,

$$u_{n,t} = (1+u_n)(u_{n+1} - u_{n-1}), \tag{1.56}$$

is very similar to the differential-difference Burgers equation,

$$u_{n,t} = (1+u_n)(u_{n+1} - u_n). \tag{1.57}$$

The former is a soliton equation and has a solution structure different from that of the differential-difference Burgers equation. □

1.4.3 The Liouville equation

It is well known that the Liouville equation,

$$\phi_{xy} = \exp\phi, \tag{1.58}$$

can also be linearized [5]. First we make the dependent variable transformation

$$\exp\phi = -2(\log f)_{xy}. \tag{1.59}$$

Since substituting this into the right-hand side of (1.58) gives

$$\phi_{xy} = -2(\log f)_{xy}, \tag{1.60}$$

we obtain, by integrating the above equation with respect to x and y,

$$\phi = -2\log f + \log g_1(x) + \log g_2(y), \tag{1.61}$$

where $g_1(x)$ and $g_2(y)$ are functions of x and y, respectively, corresponding to constants of integration. We write these using logarithms for later simplicity. Equation (1.61) is equivalent to

$$\exp\phi = g_1(x)g_2(y)/f^2, \tag{1.62}$$

and, from (1.59),

$$\exp\phi = -2(f_{xy}f - f_x f_y)/f^2. \tag{1.63}$$

Comparing these two equations, we can choose f such that

$$f_{xy} = 0, \quad f_x = g_1(x)/2, \quad f_y = g_2(y). \tag{1.64}$$

The solution for these is

$$f = u(x) + v(y), \tag{1.65}$$

where $u(x)$ and $v(y)$ are arbitrary functions in x and y, respectively. Since g_1 and g_2 are arbitrary functions, relations $f_x = u_x = g_1(x)/2$ and $f_y = v_y = g_2(y)$ always hold. Hence, an exact solution for the Liouville equation (1.58) is given by

$$\exp\phi = \frac{2u_x(x)v_y(y)}{(u(x) + v(y))^2}. \tag{1.66}$$

Remarks

(1) We have seen that the function f, in terms of which the exact solution is expressed, is written as

$$f = u(x) + v(y). \tag{1.67}$$

If we replace f by

$$f = 1 + u(x) + v(y) + \alpha u(x)v(y), \tag{1.68}$$

where α is an arbitrary constant, then $\exp\phi = -2(\log f)_{xy}$ still satisfies the Liouville equation. This is because (1.68) may be rewritten as

$$f = (1 + \alpha u(x)) \times \left(\frac{1 + u(x)}{1 + \alpha u(x)} + v(y)\right),$$

and the first factor makes no contribution to $(\log f)_{xy}$.

(2) The Liouville equation (1.58) may be transformed into the wave equation,

$$\phi_{xx} - \phi_{tt} = \exp\phi, \tag{1.69}$$

by an independent variable transformation.

(3) The following form of the Liouville equation,

$$\phi_{xy} = -2\exp 2\phi, \tag{1.70}$$

is equivalent to

$$\frac{\partial^2}{\partial x \partial y} \log q_1 = -2q_1^2, \tag{1.71}$$

through the dependent variable transformation

$$\phi = \log q_1. \tag{1.72}$$

Introducing another dependent variable,

$$q_2 = \frac{\partial}{\partial x} \log q_1, \tag{1.73}$$

(1.71) is equivalent to

$$\frac{\partial}{\partial y} q_2 = -2q_1^2. \tag{1.74}$$

Then the Liouville equation (1.70) may be written as the system of nonlinear partial differential equations

$$\begin{aligned} \frac{\partial}{\partial x} q_1 &= q_1 q_2, \\ \frac{\partial}{\partial y} q_2 &= -2q_1^2. \end{aligned} \tag{1.75}$$

The two-wave interaction equations discussed in Section 1.4.4, which are similar to the above system, can also be linearized. □

1.4.4 Two-wave interaction equations

The system of differential equations

$$\begin{aligned} \left(\frac{\partial \phi_1}{\partial t} + c_1 \frac{\partial \phi_1}{\partial x}\right) &= -\phi_1 \phi_2, \\ \left(\frac{\partial \phi_2}{\partial t} + c_2 \frac{\partial \phi_2}{\partial x}\right) &= \phi_1 \phi_2, \end{aligned} \tag{1.76}$$

describes *two-wave interactions*. This is a model which describes the interaction between prey, with population size ϕ_1 travelling at speed c_1, and predators, with population size ϕ_2 travelling at speed c_2 [6,7].

Through the independent variable transformations

$$\begin{aligned} \xi &= (x - c_2 t)/(c_1 - c_2), \\ \eta &= -(x - c_1 t)/(c_1 - c_2), \end{aligned} \tag{1.77}$$

(1.76) is transformed into the system

$$\frac{\partial \phi_1}{\partial \xi} = -\phi_1 \phi_2, \qquad \frac{\partial \phi_2}{\partial \eta} = \phi_1 \phi_2, \tag{1.78}$$

and, by the dependent variable transformation

$$\phi_1 = -\frac{\partial}{\partial \eta} \log f, \qquad \phi_2 = \frac{\partial}{\partial \xi} \log f, \tag{1.79}$$

(1.78) becomes the linear differential equation

$$\frac{\partial^2}{\partial \xi \partial \eta} f = 0. \tag{1.80}$$

The general solution of this equation is given by

$$f = u(\xi) + v(\eta), \tag{1.81}$$

in the same way as for the Liouville equation.

Remark

The Liouville equation and the two-wave interaction equation have a similar structure. In fact, a special case of the two-dimensional Toda molecule equation is equivalent to the Liouville equation, and, as shown in Chapter 4, the two-wave interaction equation is a special case of the equations generated from the Bäcklund transformation of the Toda molecule equation. Hence, the two-wave interaction equation is a soliton equation. □

Figure 1.6 illustrates how predators travelling at speed c_2 ($> c_1$) catch up with prey travelling at speed c_1, and consequently how their population size increases. Since the size of the solitary waves changes after their interaction, the situation is different from the soliton interactions described earlier. In fact, this figure shows the type of interaction typical of the soliton solutions of the Toda molecule equation.

Remark

In Figure 1.6, we have shown the solution obtained by taking

$$f = A - b_1 \tanh(p_1 \eta) + b_2 \tanh(p_2 \xi).$$

□

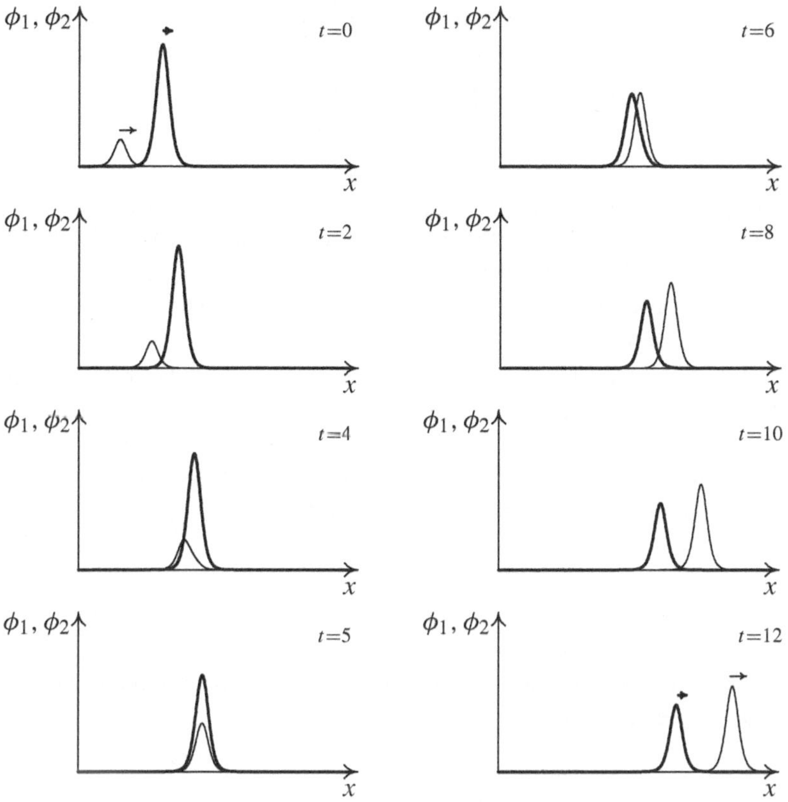

Figure 1.6. Predators travelling at speed c_2 ($> c_1$) catch up with prey travelling at speed c_1 and their population size increases. Thin and thick lines are used for the size of predators and prey populations, respectively. At $t = 0$, the leading predator comes into contact with the first of the prey. At $t = 4, 5, 6$, the predators eat the prey and increase in number. At $t = 12$, the two groups separate. Afterwards, the size of each group is unchanged.

1.5 Essentials of the direct method

In Section 1.4, we showed that certain nonlinear differential equations can be transformed into linear differential equations through a change of dependent variable. Once a nonlinear differential equation has been linearized, it is relatively easy to find an exact solution. However, only a very special class of nonlinear differential equations can be linearized. We are eager to relax this constraint and to find exact solutions for a slightly wider class of nonlinear differential equations. In this section, we explain how to transform a nonlinear differential equation into a type of a *bilinear* differential equation, often called the Hirota form, and discuss how to find an exact solution by a perturbation method. This method, through which we find solutions directly, without employing the inverse scattering method, will be referred to as the *direct method.* Outside of Japan it is called *Hirota's method.*

Remark

A *bilinear* expression is an extension of a linear expression in x_j, such as $\sum_{j=1}^{N} a_{ij}x_j$, to a second-degree expression in x_i and y_j, such as $\sum_{i,j=1}^{N} a_{ij}x_i y_j$. This may also be considered as a linear expression in x_i with y_j fixed and as a linear expression in y_j with x_i fixed. The differential equations treated below that we call bilinear do not always satisfy this property, and so it might be better to call them quadratic forms rather than bilinear forms. Nonetheless, we will always use the term bilinear. □

In order to get a feeling for what the direct method is, let us explain the fundamental ideas in relation to the KdV equation,

$$u_t + 6uu_x + u_{xxx} = 0. \tag{1.82}$$

The reader is invited to skim through the following calculations without checking the details at this point.

First of all, we look for a solution of (1.82) using the normal perturbation method. We expand u as

$$u = \varepsilon u_1 + \varepsilon^2 u_2 + \varepsilon^3 u_3 + \cdots ,$$

where ε is a small parameter, and substitute into (1.82). Collecting terms in the resulting equation at each order of ε, we have

$$\left(\frac{\partial}{\partial t} + \frac{\partial^3}{\partial x^3}\right) u_1 = 0, \tag{1.83a}$$

$$\left(\frac{\partial}{\partial t} + \frac{\partial^3}{\partial x^3}\right) u_2 = -6u_1 u_{1x}, \tag{1.83b}$$

$$\left(\frac{\partial}{\partial t} + \frac{\partial^3}{\partial x^3}\right) u_3 = -6(u_2 u_{1x} + u_1 u_{2x}), \tag{1.83c}$$

$$\vdots$$

As we remarked in Section 1.2, it is necessary to choose a solution u_1 of (1.83a) that is not a plane wave solution, but rather an exponential solution. Hence, let us choose

$$u_1 = a_1 \exp \eta, \quad \eta = Px - \Omega t, \quad \Omega = P^3, \tag{1.84}$$

where a_1 and P are arbitrary parameters. Substituting u_1 into the equation at next order (1.83b), we obtain the linear equation

$$\left(\frac{\partial}{\partial t}+\frac{\partial^3}{\partial x^3}\right)u_2=-6a_1^2P\exp 2\eta \tag{1.85}$$

for u_2, for which we may take a solution

$$u_2=a_2\exp 2\eta,\quad a_2=-a_1^2/P^2. \tag{1.86}$$

Substitution of u_2 into the next-order equation (1.83c) gives a linear equation for u_3. By this procedure, we can find u_i for $i=1,2,3,\ldots$ in succession. Hence, the solution u may be written as

$$u=\varepsilon a_1\exp\eta+\varepsilon^2a_2\exp 2\eta+\varepsilon^3a_3\exp 3\eta+\cdots, \tag{1.87}$$

the right-hand side of which diverges as $\eta\to\infty$. One idea we might use to avoid this divergence would be to express u as a ratio of polynomials G/F by employing the Padé approximation. However, there is no clear guiding principle indicating which F and G are appropriate, and it requires a great deal of effort to find them.

Instead, it seems easier to transform the dependent variable using $u=G/F$, to derive differential equations for F and G and then to find F and G as solutions to these differential equations. In fact, we see from the result in Section 1.3 that u is given by

$$u=\frac{P^2}{2}\operatorname{sech}^2\frac{\eta}{2}=\frac{2P^2\exp\eta}{(1+\exp\eta)^2}. \tag{1.88}$$

With (1.88) in mind, we will substitute $u=G/F$ into (1.82) and find F and G by a perturbation method. Substitutions of

$$\begin{aligned}
u&=\frac{G}{F},\\
u_t&=\frac{G_tF-GF_t}{F^2},\\
u_x&=\frac{G_xF-GF_x}{F^2},\\
u_{xxx}&=\frac{G_{xxx}}{F}-\frac{3G_{xx}F_x+3G_xF_{xx}+GF_{xxx}}{F^2}\\
&\quad+6\frac{G_xF_x^2+GF_{xx}F_x}{F^3}-\frac{GF_x^3}{F^4}
\end{aligned}$$

into the KdV equation yield the surprisingly complicated equation

$$\begin{aligned}
&\frac{G_t F - G F_t}{F^2} + 6\frac{G}{F}\frac{G_x F - G F_x}{F^2} \\
&+ \frac{G_{xxx}F - 3G_{xx}F_x - 3G_x F_{xx} - G F_{xxx}}{F^2} \\
&+ 6\frac{F G_x F_x^2 + F G F_{xx} F_x - G F_x^3}{F^4} = 0.
\end{aligned} \tag{1.89}$$

Snake's legs[1]

Such calculations can be easily carried out by computer, using computer algebra software. It is, however, important to understand how to manipulate formulae and calculate efficiently by hand. Otherwise, one might miss some insight into the structure of the equation. □

We will next try to separate (or *decouple*) the above complicated equation involving F and G into a simple set of equations. One possible method would be to set the term with denominator F^2 (or F^4) equal to zero. The term with denominator F^2 is given by

$$G_t F - G F_t + G_{xxx}F - 3G_{xx}F_x - 3G_x F_{xx} - G F_{xxx} = 0. \tag{1.90}$$

However, the functions F, G given by

$$\begin{aligned}
F &= (1 + \exp\eta)^2, \\
G &= 2P^2 \exp\eta,
\end{aligned} \tag{1.91}$$

which correspond to the numerator and denominator of the solitary wave solution,

$$\begin{aligned}
u &= \frac{2P^2 \exp\eta}{(1 + \exp\eta)^2}, \\
\eta &= Px - \Omega t,
\end{aligned} \tag{1.92}$$

do not satisfy (1.90). Conversely, let us look for the equation which F and G given in (1.91) *do* satisfy. Strangely enough, F and G satisfy (1.90) with the fifth term, $-3G_x F_{xx}$, replaced by $+3G_x F_{xx}$. Therefore, we change the sign of this term and transfer the remainder to the term with denominator F^4. Then

[1] Translators' note. The phrase 'Snake's legs' is a literal translation of the Japanese word 蛇足 *n.* redundancy; utter superfluousness; uselessness; coming from an ancient Chinese proverb. Since there is no equivalent usage in mathematical texts in English, the translators decided to keep the literal translation.

we can reorganize the terms in (1.89) to obtain

$$\frac{G_tF - GF_t + G_{xxx}F - 3G_{xx}F_x + 3G_xF_{xx} - GF_{xxx}}{F^2} + 6(G_xF - GF_x)\frac{GF - (FF_{xx} - F_x^2)}{F^4} = 0,$$

and we choose to adopt, as decoupled equations,

$$G_tF - GF_t + G_{xxx}F - 3G_{xx}F_x + 3G_xF_{xx} - GF_{xxx} = 0, \qquad (1.93a)$$

$$FF_{xx} - F_x^2 - GF = 0. \qquad (1.93b)$$

This is a system of bilinear (or, perhaps, quadratic) differential equations with respect to F and G and is distinctive in the pattern of derivatives. For this reason, we introduce a new binary differential operator, called the D-operator [8, 9], acting on a pair of functions $a(x)$, $b(x)$, defined by

$$D_x^n(a, b) \equiv \left(\frac{\partial}{\partial x} - \frac{\partial}{\partial y}\right)^n a(x)b(y)\bigg|_{y=x} = \frac{\partial^n}{\partial y^n} a(x+y)b(x-y)\bigg|_{y=0}, \qquad (1.94)$$

$$D_t^m D_x^n(a, b) \equiv \frac{\partial^m}{\partial s^m}\frac{\partial^n}{\partial y^n} a(t+s, x+y)b(t-s, x-y)\bigg|_{s=0,\, y=0}, \qquad (1.95)$$

where $m, n = 0, 1, 2, 3, \ldots$. Since it is rather cumbersome to write this binary operator as $D_x^n(a, b)$, instead we use the abbreviated notation

$$D_x^n(a, b) \equiv D_x^n a \cdot b. \qquad (1.96)$$

D-operators are also called *Hirota derivatives*.

Using this notation, we have

$$D_tG \cdot F = G_tF - GF_t,$$
$$D_xG \cdot F = G_xF - GF_x,$$
$$D_x^3G \cdot F = G_{xxx}F - 3G_{xx}F_x + 3G_xF_{xx} - GF_{xxx},$$
$$D_x^2F \cdot F = 2(F_{xx}F - F_x^2),$$

from which (1.93a) and (1.93b) may be concisely rewritten as

$$(D_t + D_x^3)G \cdot F = 0, \qquad (1.97a)$$

$$D_x^2F \cdot F - 2GF = 0. \qquad (1.97b)$$

Alternatively, the KdV equation,

$$u_t + 6uu_x + u_{xxx} = 0, \qquad (1.82\ bis)$$

can be transformed, through the dependent variable transformation,

$$u = 2(\log f)_{xx}, \tag{1.98}$$

into

$$\frac{\partial}{\partial x}[(f_{xt}f - f_x f_t + f_{xxxx}f - 4f_{xxx}f_x + 3f_{xx}^2)/f^2] = 0, \tag{1.99}$$

from which we obtain the bilinear equation

$$f_{xt}f - f_x f_t + f_{xxxx}f - 4f_{xxx}f_x + 3f_{xx}^2 = cf^2, \tag{1.100}$$

where c is a constant of integration [10]. Equation (1.100), with $c = 0$, may also written concisely in terms of D-operators as

$$(D_x D_t + D_x^4)f \cdot f = 0. \tag{1.101}$$

Remarks

(1) The constant c can be chosen to be zero when seeking a solitary wave (or soliton) solution.

(2) The operators on the left-hand side of (1.101) may be factorized so that it may also be written as

$$D_x(D_t + D_x^3)f \cdot f = 0. \tag{1.101$'$}$$

(3) Equations (1.97a), (1.97b) and (1.101) are called *Hirota bilinear forms.*

(4) We have employed two dependent variable transformations for u,

$$u = G/F, \quad u = 2(\log f)_{xx}.$$

From this, we might assume that the relations

$$F = f^2, \quad G = 2(f_{xx}f - f_x^2) \tag{1.102}$$

hold. It is not clear, however, that F and G defined in this way solve (1.97a) and (1.97b). □

In the next section we will present, in detail, formulae involving the new differential operator D. For now, we list only those formulae necessary to solve the bilinear equation

$$D_x(D_t + D_x^3)f \cdot f = 0 \tag{1.103}$$

by a perturbation method. We have

$$D_x D_t a \cdot 1 = a_{xt} = D_x D_t 1 \cdot a, \tag{1.104a}$$

$$D_x^4 a \cdot 1 = a_{xxxx} = D_x^4 1 \cdot a, \tag{1.104b}$$

$$D_x^m D_t^n \exp\eta_1 \cdot \exp\eta_2 = (P_1 - P_2)^m (\Omega_1 - \Omega_2)^n \exp(\eta_1 + \eta_2), \tag{1.104c}$$

where $\eta_i = P_i x + \Omega_i t + \eta_i^0$.

As in the standard perturbation method, we expand f as a power series in a small parameter ε:

$$f = 1 + \varepsilon f_1 + \varepsilon^2 f_2 + \varepsilon^3 f_3 + \cdots. \tag{1.105}$$

Substituting this expansion into (1.103) and collecting terms of each order of ε, we obtain

$$\varepsilon : D_x(D_t + D_x^3)(f_1 \cdot 1 + 1 \cdot f_1) = 0,$$

$$\varepsilon^2 : D_x(D_t + D_x^3)(f_2 \cdot 1 + f_1 \cdot f_1 + 1 \cdot f_2) = 0,$$

$$\varepsilon^3 : D_x(D_t + D_x^3)(f_3 \cdot 1 + f_2 \cdot f_1 + f_1 \cdot f_2 + 1 \cdot f_3) = 0,$$

$$\cdots$$

Using (1.104a)–(1.104b), the coefficient of ε is equivalent to

$$\frac{\partial}{\partial x}\left(\frac{\partial}{\partial t} + \frac{\partial^3}{\partial x^3}\right) f_1 = 0, \tag{1.106}$$

a linear differential equation for f_1. The solution corresponding to a solitary wave (one-soliton) is given by

$$f_1 = \exp\eta_1, \tag{1.107}$$

where $\eta_1 = P_1 x + \Omega_1 t + \eta_1^0$, and $\Omega_1 + P_1^3 = 0$.

The coefficient of ε^2 may be rearranged to give

$$2\frac{\partial}{\partial x}\left(\frac{\partial}{\partial t} + \frac{\partial^3}{\partial x^3}\right) f_2 = -D_x(D_t + D_x^3) f_1 \cdot f_1. \tag{1.108}$$

By using the property (1.104c) of D-operators, substitution of $f_1 = \exp\eta_1$ gives zero on the right-hand side of (1.108). Therefore, we are able to choose $f_2 = 0$.

This shows that the expansion of f may be truncated as the finite sum

$$f = 1 + \varepsilon f_1. \tag{1.109}$$

Since the perturbation parameter ε can be absorbed into the phase constant η_1^0 in the exponent η_1, we can conclude that

$$f = 1 + \exp \eta_1 \tag{1.110}$$

gives an exact solution of the bilinear equation. Further, this is seen to give the one-soliton solution, since

$$u = 2[\log(1 + \exp \eta_1)]_{xx} = \frac{2P_1^2 \exp \eta_1}{(1 + \exp \eta_1)^2} = \frac{P_1^2}{2} \operatorname{sech}^2 \frac{\eta_1}{2}. \tag{1.111}$$

Remarks

(1) The one-soliton solution for the KdV equation has two arbitrary parameters, η_1^0 and P_1. The parameter η_1^0 determines the position of the soliton, and P_1 its amplitude.

(2) Since the new differential operators $D_x^m D_t^n$ have the property

$$D_x^m D_t^n \exp \eta_1 \cdot \exp \eta_1 = 0, \tag{1.112}$$

perturbation expansions may be truncated as finite sums and so exact solutions are obtained. This is an important difference between bilinear and normal differential equations. If the Hirota derivative were replaced by a normal derivative, the right-hand side of (1.112) would not be zero. Then it would not be possible to choose the second term of perturbation expansion f_2 equal to zero; rather, it would be determined as a solution to a higher-order linear differential equation. Substitution of f_2 into the third approximation equation determines nonzero f_3, and so on. From this, we see that f would be given by an infinite power series.

(3) In order to find a two-soliton solution (a solution having four arbitrary parameters), we use the linear superposition principle for the solution f_1. That is, we take the solution

$$f_1 = \exp \eta_1 + \exp \eta_2, \tag{1.113}$$

where $\eta_i = P_i x + \Omega_i t + \eta_i^0$, $\Omega_i + P_i^3 = 0$ for $i = 1, 2$. Carrying out the perturbation procedure, the second term f_2 is not zero in this case, but the third term f_3 is. The exact solution is given by

$$f = 1 + \varepsilon(\exp \eta_1 + \exp \eta_2) + \varepsilon^2 a_{12} \exp(\eta_1 + \eta_2). \tag{1.114}$$

The solution $u = 2(\log f)_{xx}$ obtained from this f describes the interaction of two solitons. □

Snake's legs

The basic ideas of the direct method, as applied to the KdV, equation have been presented for the benefit of those who wish to grasp the essentials immediately. This is done because the author has often been unable to gather such information from mathematics books without reading them thoroughly. □

1.6 The *D*-operator, a new differential operator

In this section, we explain in detail the features of the D-operator introduced in Section 1.5. The D-operator is defined by

$$D_t^m D_x^n a(t,x)\cdot b(t,x) = \frac{\partial^m}{\partial s^m}\frac{\partial^n}{\partial y^n}\, a(t+s,x+y)b(t-s,x-y)|_{s=0,y=0}\,, \quad m,n=0,1,2,3,\ldots. \tag{1.115}$$

For the sake of comparison, the Leibniz rule for differentiation of a product of functions may be written in a similar way as

$$\frac{\partial^m}{\partial t^m}\frac{\partial^n}{\partial x^n}a(t,x)b(t,x) = \frac{\partial^m}{\partial s^m}\frac{\partial^n}{\partial y^n}\, a(t+s,x+y)b(t+s,x+y)|_{s=0,y=0}\,, \quad m,n=0,1,2,3,\ldots. \tag{1.116}$$

Remarks

(1) An operator O which acts on a pair of functions f and g is called a *binary operator*, and it is written as

$$O(f,g).$$

Since the D-operator is binary, it is to be written formally as

$$D_x^n(a(t,x),b(t,x)).$$

However, we use it so frequently that it is cumbersome to write the parentheses (,), and we have come to write it in the following abbreviated form:

$$D_x^n a(t,x)\cdot b(t,x).$$

(2) In the above definition of the D-operator, we have assumed that m and n are positive integers $0,1,2,\ldots$. However, it remains an open problem to extend the definition to the case where m and n are not positive integers. □

Since it is not easy to understand the nature of the D-operator from its definition alone, let us illustrate it with some simple examples:

$$\begin{aligned} D_x a \cdot b &= a_x b - a b_x, \\ D_x^2 a \cdot b &= a_{xx} b - 2a_x b_x + a b_{xx}, \\ D_x^3 a \cdot b &= a_{xxx} b - 3a_{xx} b_x + 3a_x b_{xx} - a b_{xxx}. \end{aligned} \tag{1.117}$$

For comparison, the corresponding normal derivatives of a product of functions are

$$\begin{aligned} \partial_x a \cdot b &= a_x b + a b_x, \\ \partial_x^2 a \cdot b &= a_{xx} b + 2a_x b_x + a b_{xx}, \\ \partial_x^3 a \cdot b &= a_{xxx} b + 3a_{xx} b_x + 3a_x b_{xx} + a b_{xxx}, \end{aligned} \tag{1.118}$$

where we have used an abbreviated notation $\partial_x^n \equiv \partial^n/\partial x^n$.

Remark

It is useful to remember that the formulae for D-operators in terms of derivatives are almost the same as those for normal derivatives of products. The only difference is that the signs of terms having an odd number of derivatives on the second function are negative. □

From the definition, we have

$$\begin{aligned} D_t^m D_x^n a \cdot b &= D_x^n D_t^m a \cdot b = D_x^{n-1} D_t D_x a \cdot b, \\ D_t^m D_x^n a \cdot 1 &= \partial_t^m \partial_x^n a. \end{aligned} \tag{1.119}$$

Employing the binomial expansion

$$(D_t + \varepsilon D_x)^n = D_t^n + n\varepsilon D_t^{n-1} D_x + \frac{n(n-1)}{2} \varepsilon^2 D_t^{n-2} D_x^2 + \cdots + \varepsilon^n D_x^n,$$

we obtain, because of the linearity of differential operators,

$$(D_t + \varepsilon D_x)^n a \cdot b = D_t^n a \cdot b + n\varepsilon D_t^{n-1} D_x a \cdot b + \cdots + \varepsilon^n D_x^n a \cdot b. \tag{1.120}$$

From this, we can calculate a product of D-operators, for example $3D_t D_x^2 a \cdot b$, as the coefficient of ε^2 in $(D_t + \varepsilon D_x)^3 a \cdot b$.

We define the D-operator D_z and the differential operator ∂_z by

$$\begin{aligned} D_z &= D_t + \varepsilon D_x, \\ \partial_z &= \partial_t + \varepsilon \partial_x. \end{aligned} \tag{1.121}$$

Interchanging the functions $a(t,x)$ and $b(t,x)$, we have

$$D_z^n b\cdot a = (-1)^n D_z^n a\cdot b, \tag{1.122}$$

from which we see that, if n is odd,

$$D_z^n a\cdot a = 0. \tag{1.123}$$

The obvious difference between D-operators and normal derivatives is that the action of D-operators frequently gives zero as the result. If we regard z as a new independent variable, it is only when the product of the functions $a(z)$, $b(z)$ is constant that normal differentiation ∂_z of the product gives zero. This fact implies that there are more types of solutions of Hirota equations than of normal differential equations. Note that

$$D_z a\cdot b = 0 \iff a = \text{constant}\times b. \tag{1.124}$$

The identity

$$D_z(D_z a\cdot b)\cdot c + D_z(D_z b\cdot c)\cdot a + D_z(D_z c\cdot a)\cdot b = 0 \tag{1.125}$$

holds for any functions a, b and c. Writing $D_z a\cdot b$ as $[a,b]$, we see that (1.125) can be written as the Jacobi identity

$$[[a,b],c] + [[b,c],a] + [[c,a],b] = 0, \tag{1.126}$$

which indicates one connection between the D-operator and Lie algebras.

Remark

A deep connection between bilinear equations written in terms of D-operators and Kač–Moody Lie algebras was discovered by the Kyoto group (Sato, Sato, Date, Kashiwara, Jimbo and Miwa) [11–15]. □

We may also define D-operators by the exponential identity

$$\begin{aligned}\exp(\delta D_z)a(z)\cdot b(z) &= \exp(\delta\partial_y)a(z+y)b(z-y)\big|_{y=0}\\ &= a(z+\delta)b(z-\delta),\end{aligned} \tag{1.127}$$

where δ is a parameter. If the functions $a(z)$ and $b(z)$ are continuously differentiable to all orders in z, then Taylor series expansions of $a(z+\delta)$ and $b(z-\delta)$ in δ give

$$\begin{aligned}&(1+\delta D_x+\tfrac{1}{2}\delta^2 D_x^2+\tfrac{1}{6}\delta^3 D_x^3+\cdots)a(x)\cdot b(x)\\ &\quad = [a+\delta a_x+\tfrac{1}{2}\delta^2 a_{xx}+\cdots][b-\delta b_x+\tfrac{1}{2}\delta^2 b_{xx}-\cdots].\end{aligned} \tag{1.128}$$

From each coefficient of δ^n in (1.128), we obtain expansion formulae for $D_x^n a \cdot b$ such as those shown in (1.117).

Let us next investigate the formula

$$\begin{aligned} D_z^n a \cdot b &= D_z^{n-1} D_z a \cdot b \\ &= D_z^{n-1}(a_z \cdot b - a \cdot b_z), \end{aligned} \tag{1.129}$$

in which D_z^n is expressed in terms of D_z^{n-1}. The recursive definition given by (1.129) is appropriate for use in computer algebra software to define the operator D_z^n for every n. For example, in the case of the computer algebra language REDUCE [16],

```
Operator D;
forall a,b,n,z such that fixp(n) and n>=0 let
D(a,b,z,n) = if n>0 then
             D(DF(a,z),b,z,n-1)-D(a,DF(b,z),z,n-1)
                                                  else a*b;
```

defines the operator D_z^n. Putting $a = b$ in the case of n even, we have

$$\begin{aligned} D_z^n a \cdot a &= D_z^{n-1} D_z a \cdot a \\ &= D_z^{n-1}(a_z \cdot a - a \cdot a_z) \\ &= 2D_z^{n-1} a_z \cdot a. \end{aligned} \tag{1.130}$$

For an exponential function, the relation

$$D_z^n \exp(p_1 z) \cdot \exp(p_2 z) = (p_1 - p_2)^n \exp[(p_1 + p_2)z] \tag{1.131}$$

holds. In the case of a normal derivative, we have

$$\partial_z^n \exp(p_1 z) \cdot \exp(p_2 z) = (p_1 + p_2)^n \exp[(p_1 + p_2)z], \tag{1.132}$$

from which we obtain

$$D_z^n \exp(p_1 z) \cdot \exp(p_2 z) = \frac{(p_1 - p_2)^n}{(p_1 + p_2)^n} \partial_z^n \exp(p_1 z) \cdot \exp(p_2 z). \tag{1.133}$$

Generally speaking, if F is a polynomial in $D_t, D_x, \ldots$, then

$$\begin{aligned} F(D_t, D_x, \ldots) \exp \eta_1 \cdot \exp \eta_2 = &\frac{F(\omega_1 - \omega_2, P_1 - P_2, \ldots)}{F(\omega_1 + \omega_2, P_1 + P_2, \ldots)} \\ &\times F(\partial_t, \partial_x, \ldots) \exp(\eta_1 + \eta_2), \end{aligned} \tag{1.134}$$

where $\eta_i = P_i x + \omega_i t + \cdots$ and $i = 1, 2, \ldots$. This formula is employed in the expression for the two-soliton solution of the bilinear equation,

$$F(D_t, D_x, \ldots)f \cdot f = 0. \tag{1.135}$$

For two products ab and cd, we have

$$\exp(\delta D_z)ab \cdot cd = [\exp(\delta D_z)a \cdot c][\exp(\delta D_z)b \cdot d] \tag{1.136}$$

$$= [\exp(\delta D_z)a \cdot d][\exp(\delta D_z)b \cdot c]. \tag{1.137}$$

These are similar to the *exchange formula* discussed later. Although the proof only involves using the definition and the commutativity of multiplication, establishing the formulae illustrates the full power available when manipulating bilinear equations. Equation (1.136) is easily proved by the following steps:

$$\begin{aligned}
&\exp(\delta D_z)a(z)b(z) \cdot c(z)d(z) \\
&\quad = a(z+\delta)b(z+\delta)c(z-\delta)d(z-\delta) \quad \text{(definition)} \\
&\quad = a(z+\delta)c(z-\delta)b(z+\delta)d(z-\delta) \quad \text{(commutativity)} \\
&\quad = [\exp(\delta D_z)a \cdot c][\exp(\delta D_z)b \cdot d] \quad \text{(definition).}
\end{aligned}$$

From this identity it follows that:

$$D_z ab \cdot c = a_z bc + aD_z b \cdot c, \tag{1.138}$$

$$D_z ab \cdot b = a_z b^2, \tag{1.139}$$

$$D_z^2 ab \cdot c = a_{zz} bc + 2a_z D_z b \cdot c + aD_z^2 b \cdot c, \tag{1.140}$$

$$D_z^2 ab \cdot b = a_{zz} b^2 + aD_z^2 b \cdot b, \tag{1.141}$$

$$D_z^2 ab \cdot cd = (D_z^2 a \cdot c)bd + 2(D_z a \cdot c)(D_z b \cdot d) + ac(D_z^2 b \cdot d), \tag{1.142}$$

$$D_z^3 ac \cdot bc = (D_z^3 a \cdot b)c^2 + 3(D_z a \cdot b)(D_z^2 c \cdot c), \tag{1.143}$$

$$D_z^m \exp(pz)a(z) \cdot \exp(pz)b(z) = \exp(2pz)D_z^m a(z) \cdot b(z). \tag{1.144}$$

These formulae are obtained by equating terms of the same order in δ on both sides of the above exchange formula, for appropriate choices of a, b, c and d.

Remarks

(1) The Schrödinger equation,

$$\mathrm{i}\Psi_t + \Psi_{xx} - V\Psi = 0, \tag{1.145}$$

is the most fundamental equation in quantum mechanics. Using the dependent variables transformation

$$\begin{aligned}
f' &= \Psi f, \\
V &= -D_x^2 f \cdot f/f^2 = -2(\log f)_{xx},
\end{aligned}$$

it is transformed using (1.139) and (1.141) into the bilinear equation

$$(\mathrm{i}D_t + D_x^2)f' \cdot f = 0. \tag{1.146}$$

(2) Equation (1.144) implies that a solution f of the bilinear equation

$$F(D_t, D_x, \ldots)f \cdot f = 0 \tag{1.147}$$

is invariant under the transformation

$$f(x) \to \exp(ax)f(x), \tag{1.148}$$

since

$$\begin{aligned} &F(D_t, D_x, \ldots)\exp(ax)f \cdot \exp(ax)f \\ &\quad = \exp(2ax)F(D_t, D_x, \ldots)f \cdot f = 0. \end{aligned} \tag{1.149}$$

The above fact, together with the fact that the solution for the bilinear equation is invariant under the other transformations

$$f(x) \to \exp(c)f(x), \quad f(x) \to \exp(b\partial_x)f(x) = f(x+b) \tag{1.150}$$

indicates the existence of a Lie group acting on the bilinear equation, with corresponding Lie algebra spanned by ax, $b\partial_x$ and c. □

Snake's legs

The indefiniteness of the solution that comes as a consequence of this invariance had been considered as a defect of a bilinear equation until its connection with a Lie algebra was discovered. □

Next we consider *exchange formulae*, which are most useful when deriving Bäcklund transformations.[2] The simplest case is

$$\begin{aligned} &\exp(\alpha D_z)[\exp(\beta D_z)a \cdot b] \cdot [\exp(\gamma D_z)c \cdot d] \\ &\quad = \exp\left[\frac{\beta-\gamma}{2}D_z\right]\left[\exp\left[\left(\alpha + \frac{\beta+\gamma}{2}\right)D_z\right]a \cdot d\right] \\ &\qquad \cdot \left[\exp\left[\left(-\alpha + \frac{\beta+\gamma}{2}\right)D_z\right]c \cdot b\right]. \end{aligned} \tag{1.151}$$

[2] A Bäcklund transformation is a transformation between solutions of a pair of differential equations. As will be seen in Chapter 4, a Bäcklund transformation corresponds simply to an 'exchange' in the sense used here. Viewed in this way, Bäcklund transformations seem too simple to have any significance.

We note that, while a and c do not change position, the positions of b and d are exchanged.

The exchange formula (1.151) is proved using the definition and commutativity of multiplication. We have

$$\begin{aligned}
&\exp(\alpha D_z)[\exp(\beta D_z)a \cdot b] \cdot [\exp(\gamma D_z)c \cdot d] \\
&\quad = \exp(\alpha D_z)a(z+\beta)b(z-\beta) \cdot c(z+\gamma)d(z-\gamma) \\
&\quad = a(z+\alpha+\beta)b(z+\alpha-\beta)c(z-\alpha+\gamma)d(z-\alpha-\gamma) \\
&\qquad\qquad\qquad\qquad\qquad\qquad\qquad\qquad \text{(definition)} \\
&\quad = a(z+\alpha+\beta)d(z-\alpha-\gamma)c(z-\alpha+\gamma)b(z+\alpha-\beta) \\
&\qquad\qquad\qquad\qquad\qquad\qquad\qquad\qquad \text{(commutativity)},
\end{aligned}$$

and, on the other hand,

$$\begin{aligned}
&\exp(c_1 D_z)[\exp(c_2 D_z)a \cdot d] \cdot [\exp(c_3 D_z)c \cdot b] \\
&\quad = a(z+c_1+c_2)d(z+c_1-c_2)c(z-c_1+c_3) \\
&\qquad\qquad \times b(z-c_1-c_3) \quad \text{(definition)}.
\end{aligned} \tag{1.152}$$

Necessary conditions for the exchange formula to hold are

$$\begin{aligned}
c_1 + c_2 &= \alpha + \beta, \\
c_1 - c_2 &= -\alpha - \gamma, \\
-c_1 + c_3 &= -\alpha + \gamma, \\
-c_1 - c_3 &= \alpha - \beta,
\end{aligned}$$

and so c_1, c_2 and c_3 are given by

$$\begin{aligned}
c_1 &= (\beta - \gamma)/2, \\
c_2 &= \alpha + (\beta + \gamma)/2, \\
c_3 &= -\alpha + (\beta + \gamma)/2,
\end{aligned}$$

from which the exchange formula follows.

The exchange formula (1.151) may be extended, by using the linearity of differential operators, to give

$$\begin{aligned}
&\exp(D_1)[\exp(D_2)a \cdot b] \cdot [\exp(D_3)c \cdot d] \\
&\quad = \exp\left(\frac{D_2 - D_3}{2}\right)\left[\exp\left(D_1 + \frac{D_2 + D_3}{2}\right)a \cdot d\right] \\
&\qquad \cdot \left[\exp\left(-D_1 + \frac{D_2 + D_3}{2}\right)c \cdot b\right],
\end{aligned}$$

where the D_i ($i = 1, 2, 3$) are linear combinations of D_t, D_x, D_y,

$$D_i = \alpha_i D_t + \beta_i D_x + \gamma_i D_y. \tag{1.153}$$

The following identities for the D-operator also hold:

(i) For arbitrary functions a, b, c and d,

$$\begin{aligned}&[\sinh(\delta D_z)a \cdot b][\exp(\delta D_z)c \cdot d]\\&\quad + [\sinh(\delta D_z)b \cdot c][\exp(\delta D_z)a \cdot d]\\&\quad + [\sinh(\delta D_z)c \cdot a][\exp(\delta D_z)b \cdot d] = 0.\end{aligned}$$

The coefficient of δ in the expansion of this formula with respect to δ gives the identity

$$(D_z a \cdot b)c + (D_z b \cdot c)a + (D_z c \cdot a)b = 0. \tag{1.154}$$

(ii) For arbitrary functions a, b, c and d,

$$\begin{aligned}&\sinh(\delta D_z)[\sinh(\delta D_z)a \cdot b] \cdot [\cosh(\delta D_z)c \cdot d]\\&\quad + \sinh(\delta D_z)[\sinh(\delta D_z)b \cdot c] \cdot [\cosh(\delta D_z)a \cdot d]\\&\quad + \sinh(\delta D_z)[\sinh(\delta D_z)c \cdot a] \cdot [\cosh(\delta D_z)b \cdot d] = 0.\end{aligned}$$

The coefficient of δ^2 in the expansion of this formula with respect to δ gives the Jacobi identity mentioned earlier:

$$D_z(D_z a \cdot b) \cdot c + D_z(D_z b \cdot c) \cdot a + D_z(D_z c \cdot a) \cdot b = 0. \tag{1.155}$$

Next we consider an extension [17] of the D-operator. The D-operator is based on the derivative of the transformation $u = a/b$,

$$\frac{\partial}{\partial x}\left(\frac{a}{b}\right) = \frac{D_x a \cdot b}{b^2}. \tag{1.156}$$

We can use this formula to prove the property $D_x a \cdot b = 0$ if $a = b$. Generalizing the transformation $u = a/b$ allows one to extend the D-operator in a natural way. We consider the transformation

$$u = \frac{a^m}{b^n}, \tag{1.157}$$

where m, n are not necessarily integers.

Remark

When using the direct method to find an exact solution, the basic assumption is that the solution u can be transformed as $u = g/f$ (or $u = \log(f/g)$, etc.)

and that f and g can be written as finite series. If u is written in the above form, then $g(=a^m)$ and $f(=b^n)$ are not necessarily finite series, even if a, b are themselves finite series. □

The derivative of u with respect to x can be written as

$$\frac{\partial}{\partial x}\left(\frac{a^m}{b^n}\right)=\frac{a^m}{b^n}\frac{ma_x b-nab_x}{ab}. \tag{1.158}$$

We introduce a new operator $D_{m,n,x}$ given by

$$\begin{aligned} D^j_{m,n,x}a(x)\cdot b(x)&=\left(m\frac{\partial}{\partial x}-n\frac{\partial}{\partial x'}\right)^j a(x)b(x')\bigg|_{x'=x}\\ &=\frac{\partial^j}{\partial s^j}a(x+ms)b(x-ns)\bigg|_{s=0}. \end{aligned} \tag{1.159}$$

Employing this notation, we have

$$\frac{\partial}{\partial x}\left(\frac{a^m}{b^n}\right)=\frac{a^m}{b^n}\frac{D_{m,n,x}a\cdot b}{ab}. \tag{1.160}$$

The following soliton equations are known examples which are transformed into bilinear forms by means of this operator.

The *Ginzburg–Landau equation*

$$\mathrm{i}\Psi_t+p\Psi_{xx}+q|\Psi|^2\Psi=\mathrm{i}\gamma\Psi \tag{1.161}$$

is transformed by writing

$$\Psi(x,t)=\exp[\mathrm{i}(kx-\Omega t)]G(x,t)/F^n(x,t), \tag{1.162}$$

where k, Ω and F are real and n is a complex parameter satisfying $n+n^*=2$. Substituting this gives the bilinear form:

$$\begin{aligned} &(\Omega-pk^2-\lambda+\mathrm{i}D_{n,t}+2\mathrm{i}kpD_{n,x}+pD^2_{n,x})G\cdot F=HG,\\ &\left[(1/2)p(n+n^2)D_x^2+\mathrm{i}\gamma-\lambda\right]F\cdot F-q|G|^2=HF. \end{aligned} \tag{1.163}$$

In the above, λ is an arbitrary constant and H is an arbitrary function introduced in order to decouple the equation. The D-operators $D_{1,n,t}$, $D_{1,n,x}$, $D_{1,1,x}$, using the above notation, are written as $D_{n,t}$, $D_{n,x}$, D_x. Nozaki and Bekki [18] used the bilinear form of the Ginzburg–Landau equation to derive a solitary wave solution, a hole solution, a shock wave solution and an exact solution describing a collision of two shock waves.

The *modified Kaup–Kuperschmidt equation*

$$v_t - 5(v_x v_{3x} + v_{xx}^2 + v_x^3 + 4vv_x v_{xx} + v^2 v_{3x} - v^4 v_x) + v_{5x} = 0 \tag{1.164}$$

is transformed, by writing

$$v = -3[\log(f/h^2)]_x, \tag{1.165}$$

into the bilinear form

$$\begin{aligned}(D_{2,t} + \tfrac{1}{6}D_{2,x}^5)f \cdot h = 0,\\ D_{2,x}^2 f \cdot h = 0,\end{aligned} \tag{1.166}$$

where we have used an abbreviated notation $D_{2,x}$ in place of $D_{1,2,x}$. Elimination of f from the above bilinear equations gives the Sawada–Kotera equation [19],

$$D_x(D_t + D_x^5)h \cdot h = 0, \tag{1.167}$$

and elimination of h gives the Kaup–Kuperschmidt equation,

$$\begin{aligned}D_x(D_t + \tfrac{1}{6}D_x^5)f \cdot f + \tfrac{15}{16}D_x^2 f \cdot g = 0,\\ D_x^4 f \cdot f = fg.\end{aligned} \tag{1.168}$$

Hence, the modified Kaup–Kuperschmidt equation is a Bäcklund transformation which connects the solutions of the Sawada–Kotera equation and the Kaup–Kuperschmidt equation. There exist N-soliton solutions for each of the above equations.

The nonlinear partial differential equation

$$\begin{aligned}E_t &= P,\\ P_x &= -E(P - N),\\ N_x &= 2EP,\end{aligned} \tag{1.169}$$

is transformed, using

$$\begin{aligned}\phi = \log(f/h^2), \quad \rho = \log(fh^4),\\ E = \phi_x, \quad P = \phi_{tx}, \quad N = -\rho_{tx} + 3,\end{aligned} \tag{1.170}$$

into bilinear form [20–22],

$$\begin{aligned}D_{2,x}(D_{2,x}D_{2,t} - 6)f \cdot f = 0,\\ D_{2,x}^2 f \cdot h = 0,\end{aligned} \tag{1.171}$$

which also possess an N-soliton solution.

Remark

Although it does not have a direct connection with soliton equations, if we introduce a new operator D_{xy}^n similar to the D-operator, we can obtain a unified expression for the Poisson bracket in analytical mechanics and the Hessian in differential geometry. Similar to the definition of D_x^n, we define D_{xy}^n by

$$D_{xy}^n f \cdot g \equiv \left(\frac{\partial^2}{\partial x \partial y'} - \frac{\partial^2}{\partial x' \partial y} \right)^n f(x, y) g(x', y') \bigg|_{x'=x, y'=y} . \tag{1.172}$$

This operator gives the Poisson bracket for $n = 1$:

$$\begin{aligned} [f, g]_{p,q} &\equiv \frac{\partial f}{\partial q} \frac{\partial g}{\partial p} - \frac{\partial f}{\partial p} \frac{\partial g}{\partial q} \\ &= D_{qp} f \cdot g. \end{aligned} \tag{1.173}$$

For $n = 2$,

$$D_{xy}^2 f \cdot g = f_{xx} g_{yy} - 2 f_{xy} g_{xy} + f_{yy} g_{xx},$$

which can be used to express the Hessian:

$$\text{Hessian} \equiv \begin{vmatrix} f_{xx} & f_{xy} \\ f_{yx} & f_{yy} \end{vmatrix} = \frac{1}{2} D_{xy}^2 f \cdot f. \tag{1.174}$$

□

1.7 Bilinearization of nonlinear differential equations

In this section we describe procedures for transforming nonlinear partial differential equations,

$$L(u, u_t, u_x, u_{tt}, u_{xx}, \dots) = 0, \tag{1.175}$$

into bilinear forms (or, in general, homogeneous expressions) [20–22]. There are many types of dependent variable transformations, but the most typical examples are rational, logarithmic and bi-logarithmic transformations.

1.7.1 Rational transformation

The transformation of the solution u of a nonlinear partial differential equation as

$$u = a/b$$

is called a *rational transformation*. It is the most fundamental transformation. If we employ this transformation for a differential equation whose nonlinear terms are expressed as polynomials in $u, u_x, \ldots$, we obtain a homogeneous expression with respect to a, b which is, in general, of more than fourth degree. If an appropriate way of decoupling this expression is found, it reduces to a set of quadratic forms. However, in many cases where the expression cannot be decoupled into quadratic forms, we can resort to brute force (or computer algebra) and find an exact solution, using the perturbation method, to the homogeneous expression as it stands. Often we can then deduce a quadratic form from this result.

A fundamental formula that can be used to obtain this homogeneous expression through a rational transformation is

$$\exp\left(\delta\frac{\partial}{\partial z}\right)\frac{a}{b} = \frac{\exp(\delta D_z)a\cdot b}{\cosh(\delta D_z)b\cdot b}. \tag{1.176}$$

This formula is almost obvious and can be proved directly from the definition. We have

$$\begin{aligned}
\text{LHS} &= a(z+\delta)/b(z+\delta),\\
\text{numerator of RHS} &= a(z+\delta)b(z-\delta),\\
\text{denominator of RHS} &= \tfrac{1}{2}[b(z+\delta)b(z-\delta)+b(z-\delta)b(z+\delta)]\\
&= b(z+\delta)b(z-\delta),
\end{aligned}$$

from which it follows that LHS = RHS. Expanding both sides of (1.176) with respect to the parameter δ, we have

$$\begin{aligned}
&\left(1+\delta\frac{\partial}{\partial z}+\frac{\delta^2}{2}\frac{\partial^2}{\partial z^2}+\frac{\delta^3}{6}\frac{\partial^3}{\partial z^3}+\cdots\right)\frac{a}{b}\\
&\quad= \frac{[1+\delta D_z+\frac{1}{2}\delta^2 D_z^2+\frac{1}{6}\delta^3 D_z^3+\cdots]a\cdot b}{[1+\frac{1}{2}\delta^2 D_z^2+\frac{1}{24}\delta^4 D_z^4+\cdots]b\cdot b}\\
&\quad= \left[\frac{a}{b}+\delta\frac{D_z a\cdot b}{b^2}+\frac{\delta^2}{2}\frac{D_z a\cdot b}{b^2}+\cdots\right]\\
&\qquad\times\left[1+\frac{\delta^2}{2}\frac{D_z^2 b\cdot b}{b^2}+\frac{\delta^4}{24}\frac{D_z^4 b\cdot b}{b^2}+\cdots\right]^{-1}.
\end{aligned}$$

Expanding the denominator using $(1+X)^{-1} = 1 - X + X^2 - \cdots$, and collecting terms in powers of δ, we obtain formulae which express derivatives of

$u = a/b$ in terms of the D-operator:

$$\begin{aligned}
\frac{\partial}{\partial z}\frac{a}{b} &= \frac{D_z a \cdot b}{b^2},\\
\frac{\partial^2}{\partial z^2}\frac{a}{b} &= \frac{D_z^2 a \cdot b}{b^2} - \frac{a}{b}\frac{D_z^2 b \cdot b}{b^2},\\
\frac{\partial^3}{\partial z^3}\frac{a}{b} &= \frac{D_z^3 a \cdot b}{b^2} - 3\frac{D_z a \cdot b}{b^2}\frac{D_z^2 b \cdot b}{b^2},\\
&\cdots
\end{aligned} \tag{1.177}$$

Many equations are transformed into bilinear form through rational transformations. Typical examples are given below.

The KdV equation Earlier, we were able to bilinearize the *KdV equation*,

$$u_t + 6uu_x + u_{xxx} = 0, \qquad (1.82\ bis)$$

by using trial and error. However, if we employ a rational transformation, its bilinear form is obtained almost automatically. By putting $u = G/F$ and making use of (1.177), the KdV equation may be rewritten as

$$\frac{D_t G \cdot F}{F^2} + 6\frac{G}{F}\frac{D_x G \cdot F}{F^2} + \frac{D_x^3 G \cdot F}{F^2} - 3\frac{D_x G \cdot F}{F^2}\frac{D_x^2 F \cdot F}{F^2} = 0.$$

Multiplying by F^4 on both sides and reorganizing terms, we have

$$[(D_t + D_x^3)G \cdot F]F^2 + 3[D_x G \cdot F][2GF - D_x^2 F \cdot F] = 0.$$

Therefore, if we introduce an arbitrary function λ, the above equation may be decoupled into the bilinear form

$$\begin{aligned}
(D_t + D_x^3)G \cdot F &= 3\lambda D_x G \cdot F,\\
D_x^2 F \cdot F - 2GF &= \lambda F^2.
\end{aligned} \tag{1.178}$$

The operator $D_t + D_x^3$ corresponds to the linear terms in the KdV equation, $(\partial_t + \partial_x^3)u$.

We note an indefiniteness in the bilinear equations associated with the transformation $u = G/F$. If we introduce an arbitrary function h and transform F and G to $F = hF'$ and $G = hG'$, u remains invariant. On the other hand, by virtue of properties of the D-operator, we have

$$D_t hG' \cdot hF' = h^2 D_t G' \cdot F',$$
$$D_x^2 hF' \cdot hF' = h^2 D_x^2 F' \cdot F' + (D_x^2 h \cdot h) F'^2,$$
$$D_x^3 hG' \cdot hF' = h^2 D_x^3 F' \cdot F' + 3(D_x^2 h \cdot h)(D_x G' \cdot F').$$

Therefore, under the transformation $F = hF', G = hG'$, the bilinear form is invariant,

$$\begin{aligned} (D_t + D_x^3) G' \cdot F' &= 3\lambda' D_x G' \cdot F', \\ D_x^2 F' \cdot F' - 2G'F' &= \lambda' F'^2, \end{aligned} \tag{1.179}$$

where the arbitrary function λ, which is introduced for the purpose of decoupling, is transformed to

$$\lambda' = \lambda + (D_x^2 h \cdot h)/h^2. \tag{1.180}$$

Dividing the second of the bilinear equations by F^2, we have

$$\begin{aligned} \lambda &= (D_x^2 F \cdot F)/F^2 - 2u \\ &= 2[(\log F)_{xx} - u]. \end{aligned} \tag{1.181}$$

The function λ depends on F and u. In the case of solitary wave solutions, we can choose $\lambda = 0$. However, λ plays an important role when we discuss periodic solutions, that is, those having the property $u(x, t) = u(x - L, t)$, where L is the period.

The modified KdV equation The equation we obtain by replacing the nonlinear term uu_x in the KdV equation with $u^2 u_x$,

$$w_t + 6w^2 w_x + w_{xxx} = 0, \tag{1.182}$$

is a special case of the modified KdV equation (1.50) ($\alpha = 0, \beta = 6$). Using the transformation $w = G/F$, we have

$$[(D_t + D_x^3) G \cdot F] F^2 + 3[D_x G \cdot F][2G^2 - D_x^2 F \cdot F] = 0.$$

Introducing an arbitrary function λ, we obtain the bilinear form

$$\begin{aligned} (D_t + D_x^3) G \cdot F &= 3\lambda D_x G \cdot F, \\ D_x^2 F \cdot F - 2G^2 &= \lambda F^2. \end{aligned} \tag{1.183}$$

In this case as well, the operator $D_t + D_x^3$ corresponds to the linear part of the modified KdV equation, $\partial_t + \partial_x^3$.

The nonlinear Schrödinger equation This is the equation

$$\mathrm{i}\Psi_t + \Psi_{xx} + 2c|\Psi|^2\Psi = 0, \quad c = \pm 1. \tag{1.184}$$

Through the dependent variable transformation

$$\Psi = G/F,$$

where F is a real function, we obtain

$$[(\mathrm{i}D_t + D_x^2)G \cdot F]F^2 - GF[D_x^2 F \cdot F - 2c|G|^2] = 0.$$

Therefore, its bilinear form is given by

$$\begin{aligned} (\mathrm{i}D_t + D_x^2)G \cdot F &= \lambda GF, \\ D_x^2 F \cdot F - 2c|G|^2 &= \lambda F^2. \end{aligned} \tag{1.185}$$

From the second of these equations,

$$|\Psi|^2 = [(D_x^2 F \cdot F)/F^2 - \lambda]/(2c), \tag{1.186}$$

and so the choice of λ is determined by the value of $|\Psi|^2$ at $|x| = \infty$, the function F and the sign of c. These are *envelope soliton* solutions; in the case $\lambda = 0$ (and $c > 0$), Ψ is called a *bright soliton*, and in the case $\lambda = 1$ ($c < 0$) it is called a *dark soliton*.

1.7.2 Logarithmic transformation

The transformation of a dependent variable u by using the logarithm of a function f,

$$u = 2(\log f)_{xx}, \tag{1.187}$$

is called the *logarithmic transformation*. A fundamental formula related to this transformation is

$$2\cosh\left(\delta\frac{\partial}{\partial z}\right)\log f(z) = \log[\cosh(\delta D_z)f(z) \cdot f(z)]. \tag{1.188}$$

Its proof, using just the definition, is similar to earlier ones:

$$\begin{aligned} \text{LHS} &= \log f(z+\delta) + \log f(z-\delta) \\ &= \log[f(z+\delta)f(z-\delta)], \end{aligned}$$

which is equal to the right-hand side. Expanding the above formula with respect to δ and collecting terms in powers of δ, we have

$$2\frac{\partial^2}{\partial x^2}\log f = \frac{D_x^2 f\cdot f}{f^2}, \tag{1.189a}$$

$$2\frac{\partial^2}{\partial x\partial t}\log f = \frac{D_x D_t f\cdot f}{f^2}, \tag{1.189b}$$

$$2\frac{\partial^4}{\partial x^4}\log f = \frac{D_x^4 f\cdot f}{f^2} - 3\left(\frac{D_x^2 f\cdot f}{f^2}\right)^2, \tag{1.189c}$$

$$2\frac{\partial^6}{\partial x^6}\log f = \frac{D_x^6 f\cdot f}{f^2} - 15\frac{D_x^4 f\cdot f}{f^2}\frac{D_x^2 f\cdot f}{f^2} + 30\left(\frac{D_x^2 f\cdot f}{f^2}\right)^3, \tag{1.189d}$$

$$\cdots$$

A typical example of the equations that may be bilinearized through this transformation is the ubiquitous KdV equation,

$$u_t + 6uu_x + u_{xxx} = 0. \tag{1.82 bis}$$

By introducing a potential w defined by

$$u = w_x, \tag{1.190}$$

the KdV equation may be integrated to give

$$w_t + 3w_x^2 + w_{xxx} = c, \tag{1.191}$$

where c is a constant of integration. Next, we make the dependent variable transformation

$$w = 2(\log f)_x, \tag{1.192a}$$

which is equivalent to the logarithmic transformation of u,

$$u = 2(\log f)_{xx}. \tag{1.192b}$$

From above, the KdV equation gives

$$\frac{D_x D_t f\cdot f}{f^2} + 3\left(\frac{D_x^2 f\cdot f}{f^2}\right)^2 + \frac{D_x^4 f\cdot f}{f^2} - 3\left(\frac{D_x^2 f\cdot f}{f^2}\right)^2 = c. \tag{1.193}$$

The term coming from w_x^2 has a denominator that is quartic in f. This cancels with the final term, and the bilinear form [10],

$$D_x(D_t + D_x^3)f \cdot f = cf^2, \tag{1.194}$$

is obtained automatically. In the above expression, the operator $D_t + D_x^3$ corresponds to the linear part of the KdV equation, $\partial_t + \partial_x^3$.

1.7.3 Bi-logarithmic transformation

The dependent variable transformation expressed in terms of the logarithm of a/b,

$$\phi = \log(a/b), \tag{1.195}$$

is called the *bi-logarithmic transformation*. This transformation frequently appears in conjunction with another dependent variable ρ, which is written as

$$\rho = \log(ab). \tag{1.196}$$

Fundamental formulae used with this transformation are

$$2\sinh\left(\delta\frac{\partial}{\partial x}\right)\log(a/b) = \log[\exp(\delta D_x)a \cdot b] - \log[\exp(-\delta D_x)a \cdot b], \tag{1.197a}$$

$$2\cosh\left(\delta\frac{\partial}{\partial x}\right)\log(a/b) = \log[\cosh(\delta D_x)a \cdot a] - \log[\cosh(\delta D_x)b \cdot b], \tag{1.197b}$$

$$2\cosh\left(\delta\frac{\partial}{\partial x}\right)\log(ab) = \log[\exp(\delta D_x)a \cdot b] + \log[\exp(-\delta D_x)a \cdot b]. \tag{1.197c}$$

They are proved as follows. For the first formula,

$$\begin{aligned}\text{LHS} &= \log[a(x+\delta)/b(x+\delta)] - \log[a(x-\delta)/b(x-\delta)] \\ &= \log[a(x+\delta)b(x-\delta)] - \log[a(x-\delta)b(x+\delta)] \\ &= \text{RHS};\end{aligned}$$

for the second,

$$\begin{aligned}\text{LHS} &= \log[a(x+\delta)/b(x+\delta)] + \log[a(x-\delta)/b(x-\delta)]\\ &= \log[a(x+\delta)a(x-\delta)] - \log[b(x+\delta)b(x-\delta)]\\ &= \text{RHS};\end{aligned}$$

and for the third,

$$\begin{aligned}\text{LHS} &= \log[a(x+\delta)b(x+\delta)] + \log[a(x-\delta)b(x-\delta)]\\ &= \log[a(x+\delta)b(x-\delta)] + \log[a(x-\delta)b(x+\delta)]\\ &= \text{RHS}.\end{aligned}$$

Expanding these formulae with respect to δ, we obtain

$$\frac{\partial}{\partial x}\log(a/b) = \frac{D_x a\cdot b}{ab}, \tag{1.198a}$$

$$\frac{\partial^2}{\partial x^2}\log(a/b) = \frac{D_x^2 a\cdot a}{2a^2} - \frac{D_x^2 b\cdot b}{2b^2}, \tag{1.198b}$$

$$\frac{\partial^2}{\partial x^2}\log(ab) = \frac{D_x^2 a\cdot b}{ab} - \left(\frac{D_x a\cdot b}{ab}\right)^2, \tag{1.198c}$$

$$\frac{\partial^3}{\partial x^3}\log(a/b) = \frac{D_x^3 a\cdot b}{ab} - 3\frac{D_x^2 a\cdot b}{ab}\frac{D_x a\cdot b}{ab} + 2\left(\frac{D_x a\cdot b}{ab}\right)^3, \tag{1.198d}$$

$$\cdots$$

Typical examples of nonlinear partial differential equations bilinearized through bi-logarithmic transformations are the modified KdV equation [23] and the sine–Gordon equation [24].

The modified KdV equation This is

$$v_t + 6v^2 v_x + v_{xxx} = 0, \tag{1.182 \textit{bis}}$$

and one sets $v = \mathrm{i}\phi_x$ to give

$$\phi_t - 2\phi_x^3 + \phi_{xxx} = 0. \tag{1.199}$$

Under the bi-logarithmic transformation,

$$\phi = \log(a/b), \tag{1.200}$$

we have, using (1.198a)–(1.198d),

$$\frac{D_t a \cdot b}{ab} - 2\left(\frac{D_x a \cdot b}{ab}\right)^3 + \frac{D_x^3 a \cdot b}{ab} - 3\left(\frac{D_x^2 a \cdot b}{ab}\right)\left(\frac{D_x a \cdot b}{ab}\right)$$
$$+2\left(\frac{D_x a \cdot b}{ab}\right)^3 = 0. \tag{1.201}$$

In the above case, the nonlinear term ϕ_x^3 is again cancelled automatically, and the bilinear form is written as

$$\begin{aligned}(D_t + D_x^3)a \cdot b &= 3\lambda D_x a \cdot b,\\ D_x^2 a \cdot b &= \lambda ab.\end{aligned} \tag{1.202}$$

The decoupling parameter λ can be chosen to be zero when seeking solitary wave solutions.

The sine–Gordon equation We employ the bi-logarithmic transformation

$$\phi = 2\mathrm{i}\log(f/f^*), \tag{1.203}$$

for the sine–Gordon equation,

$$\phi_{xx} - \phi_{tt} = \sin\phi, \tag{1.204}$$

where f^* is the complex conjugate of f. Since, from (1.203), we have $\exp(\mathrm{i}\phi) = (f^*/f)^2$, the right-hand side of (1.204) may be rewritten as

$$\sin\phi = \frac{1}{2\mathrm{i}}\left(\frac{f^{*2}}{f^2} - \frac{f^2}{f^{*2}}\right). \tag{1.205}$$

Substituting the expressions for derivatives,

$$\begin{aligned}\phi_{xx} &= \mathrm{i}\left(\frac{D_x^2 f \cdot f}{f^2} - \frac{D_x^2 f^* \cdot f^*}{f^{*2}}\right),\\ \phi_{tt} &= \mathrm{i}\left(\frac{D_t^2 f \cdot f}{f^2} - \frac{D_t^2 f^* \cdot f^*}{f^{*2}}\right),\end{aligned} \tag{1.206}$$

into (1.204) gives

$$[\mathrm{i}(D_x^2 - D_t^2)f \cdot f - (f^{*2} - f^2)/(2\mathrm{i})]f^{*2} - [\mathrm{i}(D_x^2 - D_t^2)f^* \cdot f^*$$
$$- (f^2 - f^{*2})/(2\mathrm{i})]f^2 = 0.$$

Hence, the bilinear form is

$$(D_x^2 - D_t^2)f \cdot f - (f^2 - f^{*2})/2 = \lambda f^2, \tag{1.207}$$

together with its complex conjugate.

Under coordinate transformations,

$$\begin{aligned} \frac{\partial}{\partial x} - \frac{\partial}{\partial t} &\to \frac{\partial}{\partial x}, \\ \frac{\partial}{\partial x} + \frac{\partial}{\partial t} &\to \frac{\partial}{\partial y}, \end{aligned} \tag{1.208}$$

the sine–Gordon equation,

$$\phi_{xx} - \phi_{tt} = \sin\phi,$$

is transformed into

$$\phi_{xy} = \sin\phi. \tag{1.209}$$

The above equation is rewritten, through the bi-logarithmic transformation,

$$\phi = 2\mathrm{i}\log(f/f^*), \tag{1.210}$$

as

$$D_x D_y f \cdot f - (f^2 - f^{*2})/2 = \lambda f^2, \tag{1.211}$$

where λ is real and is chosen to be zero when seeking solitary wave solutions.

1.8 Solutions of bilinear equations

We have shown that the KdV equation,

$$u_t + 6uu_x + u_{xxx} = 0, \tag{1.82 bis}$$

is transformed by

$$u = 2(\log f)_{xx} \tag{1.187 bis}$$

into the bilinear equation,

$$D_x(D_t + D_x^3)f \cdot f = cf^2. \tag{1.194 bis}$$

In this section, we describe the perturbation method used to find its exact solution. Since we will find solitary wave solutions, we set $c = 0$.

The perturbation method consists of expanding f with respect to a small parameter ε to obtain

$$f = 1 + \varepsilon f_1 + \varepsilon^2 f_2 + \varepsilon^3 f_3 + \cdots , \qquad (1.212)$$

and then finding each coefficient f_n successively for $n = 1, 2, \ldots$. Generally, when using this method, the expansion continues to infinite order in ε, and we truncate the expansion at an appropriate finite order. Therefore, the solution we obtain is no more than an approximation. However, when performing the perturbation method for bilinear equations, an appropriate choice of f_1 (f_1 satisfies a linear differential equation) makes the infinite expansion truncate with a finite number of terms, and as a result the solution is an exact one.

Substituting the expansion formula of f into the bilinear equation and arranging it at each order of ε, we have

$$\begin{aligned}
&\varepsilon : D_x(D_t + D_x^3)(f_1 \cdot 1 + 1 \cdot f_1) = 0,\\
&\varepsilon^2 : D_x(D_t + D_x^3)(f_2 \cdot 1 + f_1 \cdot f_1 + 1 \cdot f_2) = 0,\\
&\varepsilon^3 : D_x(D_t + D_x^3)(f_3 \cdot 1 + f_2 \cdot f_1 + f_1 \cdot f_2 + 1 \cdot f_3) = 0,\\
&\varepsilon^4 : D_x(D_t + D_x^3)(f_4 \cdot 1 + f_3 \cdot f_1 + f_2 \cdot f_2 + f_1 \cdot f_3 + 1 \cdot f_4) = 0,\\
&\quad \ldots
\end{aligned}$$

The order-ε equation may be rewritten as a linear differential equation for f_1:

$$\frac{\partial}{\partial x}\left(\frac{\partial}{\partial t} + \frac{\partial^3}{\partial x^3}\right) f_1 = 0. \qquad (1.213)$$

We showed at the beginning of Section 1.5 that the solution describing a solitary wave (one-soliton) is given by

$$f_1 = \exp \eta_1, \qquad (1.107\ bis)$$

where $\eta_1 = P_1 x + \Omega_1 t + \eta_1^0$ and $\Omega_1 + P_1^3 = 0$. We begin here by finding a two-soliton solution, that is a solution describing the interaction of two solitons.

To this end, we choose the solution to the linear differential equation (1.213) to be

$$f_1 = \exp \eta_1 + \exp \eta_2, \qquad (1.214)$$

where $\eta_i = P_i x + \Omega_i t + \text{constant}$ and $\Omega_i + P_i^3 = 0$ for $i = 1, 2$. The relationship $\Omega_i + P_i^3 = 0$ is called the *nonlinear dispersion relation*. The order-ε^2

equation is

$$2\frac{\partial}{\partial x}\left(\frac{\partial}{\partial t}+\frac{\partial}{\partial x^3}\right)f_2 = -D_x(D_t+D_x^3)f_1\cdot f_1. \tag{1.215}$$

Substituting $f_1 = \exp\eta_1 + \exp\eta_2$ into the right-hand side, we have, from the property of the D-operator (1.104c),

$$\begin{aligned}
&D_x(D_t+D_x^3)f_1\cdot f_1\\
&\quad = D_x(D_t+D_x^3)(\exp\eta_1+\exp\eta_2)\cdot(\exp\eta_1+\exp\eta_2)\\
&\quad = 2D_x(D_t+D_x^3)\exp\eta_1\cdot\exp\eta_2\\
&\quad = 2(P_1-P_2)[\Omega_1-\Omega_2+(P_1-P_2)^3]\exp(\eta_1+\eta_2).
\end{aligned}$$

We may choose the solution of this to be

$$f_2 = a_{12}\exp(\eta_1+\eta_2), \tag{1.216}$$

where, using (1.104c), the coefficient a_{12} is given by

$$\begin{aligned}
a_{12} &= -\frac{2(P_1-P_2)[\Omega_1-\Omega_2+(P_1-P_2)^3]}{2(P_1+P_2)[\Omega_1+\Omega_2+(P_1+P_2)^3]}\\
&= \frac{(P_1-P_2)^2}{(P_1+P_2)^2}.
\end{aligned} \tag{1.217}$$

Substituting the expressions for f_1 and f_2 given above into the linear differential equation for f_3,

$$D_x(D_t+D_x^3)(f_3\cdot 1 + f_2\cdot f_1 + f_1\cdot f_2 + 1\cdot f_3) = 0, \tag{1.218}$$

we obtain

$$\begin{aligned}
&D_x(D_t+D_x^3)(f_2\cdot f_1+f_1\cdot f_2)\\
&\quad = 2D_x(D_t+D_x^3)\exp(\eta_1+\eta_2)\cdot(\exp\eta_1+\exp\eta_2)\\
&\quad = 2P_2(\Omega_2+P_2^3)\exp(2\eta_1+\eta_2)+2P_1(\Omega_1+P_1^3)\exp(\eta_1+2\eta_2),
\end{aligned}$$

in which the inhomogeneous term is zero by virtue of the nonlinear dispersion relation $\Omega_i + P_i^3 = 0$ for $i = 1, 2$. Hence, we may choose $f_3 = 0$. Consequently, at order ε^4 we have the linear equation

$$D_x(D_t+D_x^3)(f_4\cdot 1 + f_2\cdot f_2 + 1\cdot f_4) = 0. \tag{1.219}$$

In this case as well, the inhomogeneous term vanishes because

$$D_x(D_t + D_x^3) f_2 \cdot f_2 = 0 \tag{1.220}$$

and we may also choose $f_4 = 0$. Substituting the above results into the perturbation expansion of f, we have

$$f = 1 + \varepsilon(\exp \eta_1 + \exp \eta_2) + \varepsilon^2 a_{12} \exp(\eta_1 + \eta_2). \tag{1.221a}$$

In the above equation, ε is the small parameter giving the perturbation expansion. Since each η_i is given by

$$\eta_i = P_i x + \Omega_i t + \text{constant}, \tag{1.221b}$$

ε can be absorbed into the constant. Hence,

$$f = 1 + \exp \eta_1 + \exp \eta_2 + a_{12} \exp(\eta_1 + \eta_2) \tag{1.222}$$

gives the two-soliton solution for the KdV equation in its bilinear form. Therefore, the two-soliton solution u for the KdV equation is expressed, in terms of f, as

$$u = 2(\log f)_{xx}. \tag{1.223}$$

This result indicates that two solitons are not destroyed after their interaction. Figure 1.7 shows a plot of their interaction; it was calculated numerically using an appropriate difference scheme.

Remark

In the expression for f, a_{12} is an important quantity which determines the phase shift, that is the change of position, caused by the interaction of two solitons. Mathematically, the structure of a_{12} is one of the ingredients that determines the type of group acting on the bilinear equation. We have

$$a_{12} = -\frac{(P_1 - P_2)[\Omega_1 - \Omega_2 + (P_1 - P_2)^3]}{(P_1 + P_2)[\Omega_1 + \Omega_2 + (P_1 + P_2)^3]}, \tag{1.224}$$

whose structure clearly reflects the structure of the KdV equation in its bilinear form,

$$D_x(D_t + D_x^3) f \cdot f = 0. \tag{1.225}$$

□

We next consider a bilinear equation of the form

$$F(D_t, D_x, D_y, \ldots) f \cdot f = 0, \tag{1.226}$$

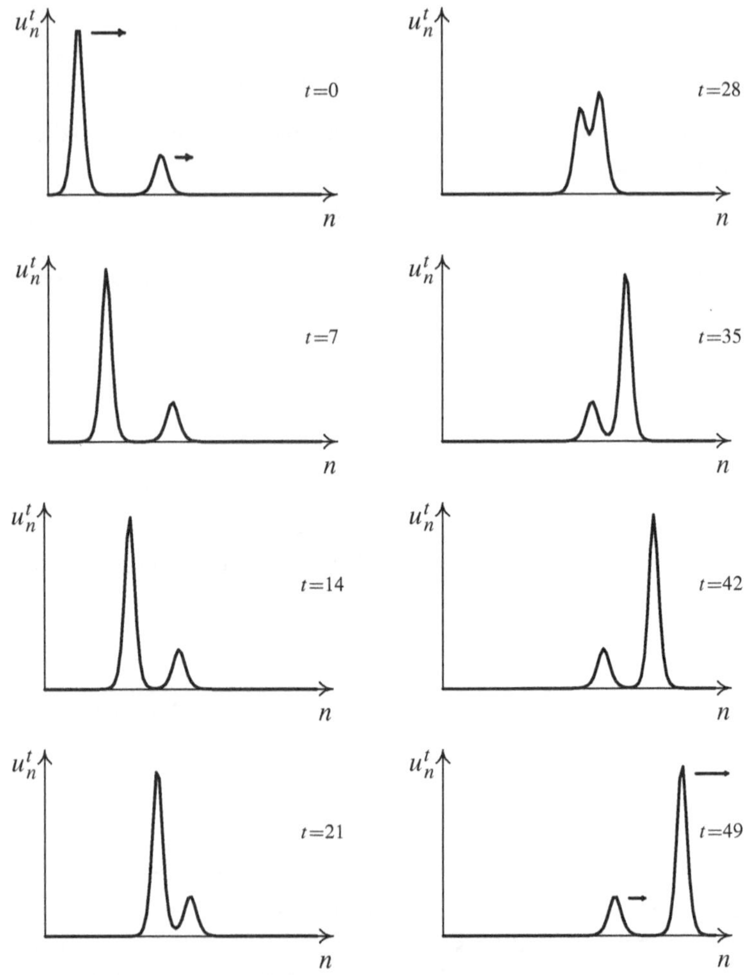

Figure 1.7. The numerical simulation of the interaction of two solitons using the fully discrete soliton equation, $u_n^{t+1} - u_n^t = \delta(u_{n-1}^{t+1} u_n^t - u_n^{t+1} u_{n+1}^t)$, where $n = \ldots, -1, 0, 1, \ldots$ and δ is the lattice spacing. At $t = 0$, a large soliton is about to interact with a small soliton. At $t = 28$ they are overlapping, and at $t = 49$ they recover their original shapes. Their shapes are not smooth because their time evolutions take place at discrete time and space steps.

where F is a general polynomial in $D_t, D_x, D_y, \ldots$. Let us call this equation a *KdV-type bilinear equation*. The distinguishing feature of a KdV-type bilinear equation is that it has just one dependent variable f. Introducing vector notation,

$$\boldsymbol{D} = (D_t, D_x, D_y, \ldots), \tag{1.227}$$

we can rewrite (1.226) as

$$F(\boldsymbol{D})f \cdot f = 0. \tag{1.228}$$

Further, we impose the following conditions on F:

$$\text{(i) } F(-\boldsymbol{D}) = F(\boldsymbol{D}), \tag{1.229a}$$

$$\text{(ii) } F(\boldsymbol{0}) = 0. \tag{1.229b}$$

Condition (i) is not essential. If a polynomial in D, F_{odd}, is an odd function, then $F_{\text{odd}}(\boldsymbol{D})f \cdot f = 0$ automatically. Condition (ii) is required in order to find a solitary wave (soliton) solution. We have expanded $f = 1 + \varepsilon f_1 + \cdots$ in performing a perturbation method. However, without this condition, the ε^0-term is given by

$$F(\boldsymbol{0})1 \cdot 1 \neq 0,$$

and therefore f is not a solution.

Remark

If condition (ii) does not hold, the solutions found are quasi-periodic. Akira Nakamura [25] has shown that the KdV-type bilinear equations possess genus-two quasi-periodic solutions in this case. □

Almost the same perturbation method, applied to the bilinear form of the KdV equation, may be used to construct the two-soliton solution to any KdV-type bilinear equation [20–22]. We have already stated that the relation

$$\begin{aligned} F(D_t, D_x, \ldots)\exp\eta_1 \cdot \exp\eta_2 = {} & \frac{F(\omega_1 - \omega_2, P_1 - P_2, \ldots)}{F(\omega_1 + \omega_2, P_1 + P_2, \ldots)} \\ & \times F(\partial_t, \partial_x, \ldots)\exp(\eta_1 + \eta_2) \end{aligned} \tag{1.230}$$

holds for any polynomial F. We can also rewrite this relation in the succinct form

$$F(\boldsymbol{D})\exp\eta_1 \cdot \exp\eta_2 = \frac{F(\boldsymbol{P}_1 - \boldsymbol{P}_2)}{F(\boldsymbol{P}_1 + \boldsymbol{P}_2)} F(\boldsymbol{\partial})\exp(\eta_1 + \eta_2), \tag{1.230'}$$

where

$$\boldsymbol{P}_1 \pm \boldsymbol{P}_2 = (\omega_1 \pm \omega_2, P_1 \pm P_2, \ldots),$$
$$\boldsymbol{\partial} = (\partial_t, \partial_x, \ldots).$$

By the perturbation method, the two-soliton solution for the KdV-type bilinear equations,

$$F(\boldsymbol{D}) f \cdot f = 0, \tag{1.231}$$

is

$$f = 1 + \exp \eta_1 + \exp \eta_2 + a_{12} \exp(\eta_1 + \eta_2), \tag{1.232}$$

where

$$\eta_i = P_i x + \Omega_i t + Q_i y + \text{constant} \quad (i = 1, 2). \tag{1.233}$$

This last equation may be written as

$$\eta_i = \boldsymbol{P}_i \cdot \boldsymbol{x}_i + \text{constant} \tag{1.234}$$

using vector notation:

$$\boldsymbol{P}_i = (\Omega_i, P_i, Q_i, \ldots),$$
$$\boldsymbol{x}_i = (t, x, y, \ldots).$$

The nonlinear dispersion relation is given by

$$F(\boldsymbol{P}_i) = 0 \quad (i = 1, 2), \tag{1.235}$$

and the phase shift a_{12} is given by

$$a_{12} = -\frac{F(\boldsymbol{P}_1 - \boldsymbol{P}_2)}{F(\boldsymbol{P}_1 + \boldsymbol{P}_2)}. \tag{1.236}$$

Remarks

(1) An extension of the KdV-type bilinear equations,

$$F(\boldsymbol{D})[f \cdot f + (L_1 f) \cdot (L_2 f)] = 0, \tag{1.237}$$

also possesses a two-soliton solution. In the above equation, L_i are linear differential operators,

$$\begin{aligned} L_i &= L_i(\boldsymbol{\partial}), \\ L_i(\boldsymbol{0}) &= 0, \end{aligned} \tag{1.238}$$

for $i = 1, 2$, and $F(\boldsymbol{D})$ satisfies the same conditions as before:

$$\text{(i) } F(-\boldsymbol{D}) = F(\boldsymbol{D}), \tag{1.229a bis}$$

$$\text{(ii) } F(\boldsymbol{0}) = 0. \tag{1.229b bis}$$

The two-soliton solution is of almost the same form as that of a KdV-type bilinear equation,

$$f = 1 + \exp\eta_1 + \exp\eta_2 + a_{12}\exp(\eta_1 + \eta_2), \tag{1.239}$$

where $\eta_i = P_i x + \Omega_i t + Q_i y + \cdots + \text{constant}$ and $F(\boldsymbol{P}_i) = 0$, for $i = 1, 2$. The only difference lies in the phase shift term, which is given by

$$\begin{aligned} a_{12} &= -\frac{F(\boldsymbol{P}_1 - \boldsymbol{P}_2)}{F(\boldsymbol{P}_1 + \boldsymbol{P}_2)}(1 + \gamma), \\ \gamma &= \frac{1}{2}[L_1(\boldsymbol{P}_1)L_2(\boldsymbol{P}_2) + L_1(\boldsymbol{P}_2)L_2(\boldsymbol{P}_1)]. \end{aligned} \tag{1.240}$$

(2) We investigate briefly equations of trilinear form which possess a three-soliton solution. The two-soliton solution for the KdV-type bilinear equation

$$F(\boldsymbol{D})f \cdot f = 0 \tag{1.241}$$

is given by

$$f = 1 + \exp\eta_1 + \exp\eta_2 + a_{12}\exp(\eta_1 + \eta_2). \tag{1.242}$$

By employing the above $F(\boldsymbol{D})$, let us consider the trilinear form:

$$F(\boldsymbol{D})[F(\boldsymbol{\partial})f] \cdot [F(\boldsymbol{D})f \cdot f] = 0. \tag{1.243}$$

This type of equation, in addition, possesses a three-soliton solution,

$$\begin{aligned} f &= 1 + \exp\eta_1 + \exp\eta_2 + \exp\eta_3 + a_{12}\exp(\eta_1 + \eta_2) \\ &\quad + a_{13}\exp(\eta_1 + \eta_3) + a_{23}\exp(\eta_2 + \eta_3) \\ &\quad + a_{12}a_{13}a_{23}\exp(\eta_1 + \eta_2 + \eta_3), \end{aligned} \tag{1.244}$$

which implies an essential difference between a bilinear form and a trilinear form. □

We now return to the bilinear form of the KdV equation,

$$D_x(D_t + D_x^3)f \cdot f = 0, \tag{1.226 bis}$$

and find a three-soliton solution. Choosing

$$f_1 = \exp \eta_1 + \exp \eta_2 + \exp \eta_3 \tag{1.245}$$

as a solution to the linear differential equation (1.213), the perturbation method gives

$$\begin{aligned} f &= 1 + \exp \eta_1 + \exp \eta_2 + \exp \eta_3 \\ &\quad + a_{12} \exp(\eta_1 + \eta_2) + a_{13} \exp(\eta_1 + \eta_3) + a_{23} \exp(\eta_2 + \eta_3) \\ &\quad + a_{123} \exp(\eta_1 + \eta_2 + \eta_3), \end{aligned} \tag{1.246}$$

where

$$\eta_i = P_i x + \Omega_i t + \text{constant}, \quad \Omega_i + P_i^3 = 0 \quad (i = 1, 2, 3),$$
$$a_{ij} = \frac{(P_i - P_j)^2}{(P_i + P_j)^2} \quad (i, j = 1, 2, 3),$$

and $a_{123} = a_{12}a_{13}a_{23}$. Although it is a complicated calculation to obtain this, it is simple when using computer algebra software.

By writing

$$a_{ij} = \exp A_{ij},$$

we may express f as

$$f = \sum \exp \left[\sum_{i=1}^{3} \mu_i \eta_i + \sum_{i<j}^{(3)} A_{ij} \mu_i \mu_j \right].$$

In this expression, the first $\sum$ means the summation over all possible combinations of $\mu_1 = 0, 1,\ \mu_2 = 0, 1,\ \mu_3 = 0, 1$. For example, the choice $\mu_1 = 1,\ \mu_2 = 0,\ \mu_3 = 0$ gives $\exp \eta_1$, the choice $\mu_1 = 1, \mu_2 = 1, \mu_3 = 0$ gives $a_{12} \exp(\eta_1 + \eta_2)$, and the choice $\mu_1 = 1, \mu_2 = 1, \mu_3 = 1$ gives $a_{123} \exp(\eta_1 + \eta_2 + \eta_3)$. The notation $\sum_{i<j}^{(3)}$ means the summation over all possible pairs (i, j) chosen from the set $\{1, 2, 3\}$, with the condition that $i < j$. Hence, f is written as

$$\begin{aligned} f &= 1 + \exp \eta_1 + \exp \eta_2 + \exp \eta_3 \\ &\quad + \exp(A_{12} + \eta_1 + \eta_2) + \exp(A_{13} + \eta_1 + \eta_3) \\ &\quad + \exp(A_{23} + \eta_2 + \eta_3) + \exp(A_{12} + A_{13} + A_{23} + \eta_1 + \eta_2 + \eta_3). \end{aligned}$$

By employing the above notation, the N-soliton solution is expressed as

$$f = \sum \exp\left[\sum_{i=1}^{N} \mu_i \eta_i + \sum_{i<j}^{(N)} A_{ij} \mu_i \mu_j\right], \tag{1.247}$$

where the first $\sum$ means a summation over all possible combinations of $\mu_1 = 0, 1,\ \mu_2 = 0, 1, \ldots, \mu_N = 0, 1$, and $\sum_{i<j}^{(N)}$ means a summation over all possible pairs (i, j) chosen from the set $\{1, 2, \ldots, N\}$, with the condition that $i < j$.

For all KdV-type bilinear equations,

$$F(\boldsymbol{D}) f \cdot f = 0, \tag{1.241 bis}$$

having N-soliton solutions, f has the form (1.247). Putting

$$\begin{aligned} \boldsymbol{P}_i &= (\Omega_i, P_i, Q_i, \ldots), \\ \boldsymbol{x}_i &= (t, x, y, \ldots), \end{aligned}$$

we have, for $i, j = 1, 2, 3, \ldots, N$,

$$\begin{aligned} \eta_i &= \boldsymbol{P}_i \cdot \boldsymbol{x}_i + \text{constant}, \\ F(\boldsymbol{P}_i) &= 0, \end{aligned} \tag{1.248}$$

and the phase shift a_{ij} is given by

$$a_{ij} = -\frac{F(\boldsymbol{P}_i - \boldsymbol{P}_j)}{F(\boldsymbol{P}_i + \boldsymbol{P}_j)}. \tag{1.249}$$

The function $F(\boldsymbol{D})$ is not arbitrary, however; it must satisfy the condition

$$\sum F\left(\sum_{i=1}^{N} \sigma_i \boldsymbol{P}_i\right) \prod_{i<j}^{(N)} F(\sigma_i \boldsymbol{P}_i - \sigma_j \boldsymbol{P}_j) \sigma_i \sigma_j = 0, \tag{1.250}$$

where $\sum$ means the summation over all possible combinations of $\sigma_1 = 0, 1,\ \sigma_2 = 0, 1, \ldots, \sigma_N = 0, 1$. This is called the Hirota condition [20–22, 26–29].

Remark

Though this condition is obtained by substituting the N-soliton solution f into the KdV-type bilinear equation, it is not easy to derive or to prove. As is shown in Chapter 2, once the structure of a bilinear equation has been clarified, we can use a completely different technique to obtain a simple proof of (1.247) for every N. Therefore we need not prove the above condition for arbitrary N. However, the search for a function F satisfying the condition was formerly the only way to discover a soliton equation by the direct method. Except for the case $F(\boldsymbol{D}) = D_1 D_2 D_3 D_4$ (J. Hietarinta, private communication), there appear to be no bilinear equations that possess a three-soliton solution but no four-soliton solution. Hietarinta checked the condition for a three-soliton solution using computer algebra software and discovered many new (potential) soliton equations [26–29]. □

Bilinear equations are transformed into normal nonlinear partial differential equations by employing the formulae described in the following section. The rest of this section is devoted to listing KdV-type bilinear equations satisfying the above condition for arbitrary N. We also give the associated dependent variable transformations and nonlinear partial differential equations. We have omitted partial difference equations from this list.

- KdV equation

$$u_t + 6uu_x + u_{xxx} = 0, \tag{1.251a}$$

$$u = 2(\log f)_{xx}, \tag{1.251b}$$

$$D_x(D_t + D_x^3) f \cdot f = 0. \tag{1.251c}$$

- Lax fifth-order KdV equation

$$u_t + 10\left(u^3 + \tfrac{1}{2}u_x^2 + uu_{xx}\right)_x + u_{xxxxx} = 0, \tag{1.252a}$$

$$u = 2(\log f)_{xx}, \tag{1.252b}$$

$$\left[D_x(D_t + D_x^5) - \tfrac{5}{3}D_s(D_s + D_x^3)\right] f \cdot f = 0, \tag{1.252c}$$

where f also satisfies the bilinear equation

$$D_x(D_s + D_x^3) f \cdot f = 0, \tag{1.253}$$

involving an auxiliary variable s.

- Sawada–Kotera equation

$$u_t + 15(u^3 + uu_{xx}) + u_{xxxxx} = 0, \tag{1.254a}$$

$$u = 2(\log f)_{xx}, \tag{1.254b}$$

$$D_x(D_t + D_x^5) f \cdot f = 0. \tag{1.254c}$$

- Boussinesq equation

$$u_{tt} - u_{xx} - 3(u^2)_{xx} - u_{xxxx} = 0, \tag{1.255a}$$

$$u = 2(\log f)_{xx}, \tag{1.255b}$$

$$(D_t^2 - D_x^2 - D_x^4) f \cdot f = 0. \tag{1.255c}$$

- Kadomtsev–Petviashvili equation (KP equation)

$$(u_t + 6uu_x + u_{xxx})_x \pm u_{yy} = 0, \tag{1.256a}$$

$$u = 2(\log f)_{xx}, \tag{1.256b}$$

$$\left[D_x(D_t + D_x^3) \pm D_y^2\right] f \cdot f = 0. \tag{1.256c}$$

- Model equations for shallow water waves

(i)
$$u_t - u_{xx} - 4uu_t + 2u_x \int_x^\infty u_t \mathrm{d}x' + u_x = 0, \tag{1.257a}$$

$$u = 2(\log f)_{xx}, \tag{1.257b}$$

$$\left[D_x(D_t - D_t D_x^2 + D_x) + \tfrac{1}{3} D_t(D_s + D_x^3)\right] f \cdot f = 0, \tag{1.257c}$$

where f also satisfies the bilinear equation

$$D_x(D_s + D_x^3) f \cdot f = 0, \tag{1.258}$$

involving an auxiliary variable s.

(ii)
$$u_t - u_{xxt} - 3uu_t + 3u_x \int_x^\infty u_t \mathrm{d}x' + u_x = 0, \tag{1.259a}$$

$$u = 2(\log f)_{xx}, \tag{1.259b}$$

$$D_x(D_t - D_t D_x^2 + D_x) f \cdot f = 0. \tag{1.259c}$$

- Toda lattice equation

$$\frac{\partial^2}{\partial t^2} \log[1 + V_n(t)] = V_{n+1}(t) - 2V_n(t) + V_{n-1}(t), \tag{1.260a}$$

$$V_n(t) = \frac{\partial^2}{\partial t^2} \log f_n(t) = \frac{f_{n+1}(t) f_{n-1}(t)}{f_n(t)^2} - 1, \tag{1.260b}$$

$$\left[D_t^2 - 4\sinh^2\left(\tfrac{1}{2} D_n\right)\right] f_n \cdot f_n = 0. \tag{1.260c}$$

1.9 Transformation from bilinear to nonlinear form

If it has been shown that a certain bilinear form satisfies the Hirota condition and so has an N-soliton solution, or if an unusual bilinear form is discovered, we would like to transform it into a normal partial differential equation. There is no unique nonlinear partial differential equation obtained in this process; for different choices of dependent variable transformations, the bilinear form can be transformed into completely different types of differential equation.

The following formulae facilitate transformation from bilinear forms to nonlinear partial differential equations.

1.9.1 Rational transformation

The fundamental formula is

$$\exp(\delta D_x)a \cdot b = \{\exp[2\cosh(\delta\partial_x)\log b]\}[\exp(\delta\partial_x)(a/b)]. \tag{1.261}$$

This is easily proved:

$$\begin{aligned}\text{RHS} &= \exp\big[\log\big(b(x+\delta)b(x-\delta)\big)\big]a(x+\delta)/b(x+\delta)\\ &= a(x+\delta)b(x-\delta)\\ &= \text{LHS}.\end{aligned}$$

Introducing dependent variables Ψ and u defined by

$$\Psi = a/b, \qquad u = 2(\log b)_{xx},$$

and expanding (1.261) with respect to δ, we obtain

$$(D_x a \cdot b)/b^2 = \Psi_x, \tag{1.262a}$$

$$(D_x^2 a \cdot b)/b^2 = \Psi_{xx} + u\Psi, \tag{1.262b}$$

$$(D_x^3 a \cdot b)/b^2 = \Psi_{xxx} + 3u\Psi_x, \tag{1.262c}$$

$$(D_x^4 a \cdot b)/b^2 = \Psi_{xxxx} + 6u\Psi_{xx} + (u_{xx} + 3u^2)\Psi, \tag{1.262d}$$

$$(D_x^5 a \cdot b)/b^2 = \Psi_{xxxxx} + 10u\Psi_{xxx} + 5(u_{xx} + 3u^2)\Psi_x. \tag{1.262e}$$

1.9.2 Logarithmic transformation

The fundamental formula is

$$\cosh(\delta D_x)f \cdot f = \exp[2\cosh(\delta\partial_x)\log f]. \tag{1.263}$$

Let us define a dependent variable u by

$$u = 2(\log f)_{xx}. \tag{1.264}$$

The Taylor expansion of (1.263) with respect to δ gives

$$(D_x^2 f \cdot f)/f^2 = u, \tag{1.265a}$$
$$(D_x^4 f \cdot f)/f^2 = u_{xx} + 3u^2, \tag{1.265b}$$
$$(D_x^6 f \cdot f)/f^2 = u_{xxxx} + 15uu_{xx} + 15u^3, \tag{1.265c}$$
$$\ldots$$

1.9.3 Bi-logarithmic transformation

The fundamental formula is

$$\exp(\delta D_x)a \cdot b = \exp[\sinh(\delta \partial_x) \log(a/b) + \cosh(\delta \partial_x) \log(ab)]. \tag{1.266}$$

Defining dependent variables ϕ and ρ by

$$\phi = \log(a/b), \quad \rho = \log(ab), \tag{1.267}$$

we have, from the Taylor expansion of (1.266),

$$(D_x a \cdot b)/ab = \phi_x, \tag{1.268a}$$
$$(D_x^2 a \cdot b)/ab = \rho_{xx} + \phi_x^2, \tag{1.268b}$$
$$(D_x^3 a \cdot b)/ab = \phi_{xxx} + 3\phi_x \rho_{xx} + \phi_x^3, \tag{1.268c}$$
$$(D_x^4 a \cdot b)/ab = \rho_{xxxx} + 4\phi_x \rho_{xxx} + 3\rho_{xx}^2 + 6\phi_x^2 \rho_{xx} + \phi_x^4, \tag{1.268d}$$
$$\ldots$$

2
Determinants and pfaffians

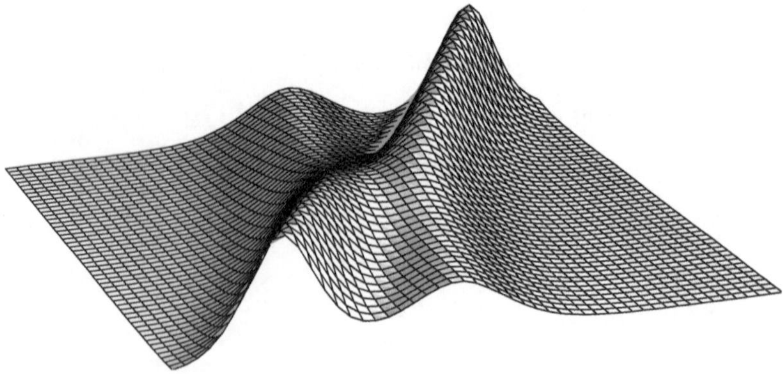

Overtaking interaction.

2.1 Introduction

In Chapter 1, we discussed transformations between soliton equations and bilinear equations and the solution of such equations. But what is a bilinear equation, or, more concretely, what mathematical structures are characteristic of bilinear equations? One answer to this question is the existence of groups (related to affine Lie algebras) which act on bilinear equations. In fact, a collective understanding of soliton equations has developed from this viewpoint, and many new soliton equations have been found using this group-theoretical method. This approach, however, calls for a deep knowledge of algebra, and, even when this has been attained, it is difficult to apply. Soon after the birth of quantum mechanics, group theory became a great craze (called *Gruppenpest*), where many people only studied group theory and never managed to apply it.

Since this book is aimed at students of science and technology, we omit most of the group theory.

A new viewpoint, discovered by Mikio Sato [12, 13], is to regard bilinear equations as equivalent to Plücker relations in a Grassmann manifold. This interpretation of soliton equations, based on a deep knowledge of mathematics, is admirable and beautiful, and has had a strong influence on the author. However, its later development has been so abstract that the author has not been able to understand it completely.

Abstract theory does not seem to be of value to pragmatic researchers, such as the author, who wish to find exact solutions to nonlinear differential equations and to discover new soliton equations. Such theory has, up to now, not been very efficient in generating nonlinear partial *difference* equations, which play an important role as difference schemes for nonlinear partial differential equations. Since the author believes that soliton equations can be discretized and that the era of difference equations has yet to come, Sato theory will not be thoroughly investigated in this book. In this chapter we consider bilinear equations arising from soliton equations from a unified viewpoint, as nothing more than pfaffian identities.

A particular pfaffian identity can be expressed symbolically as

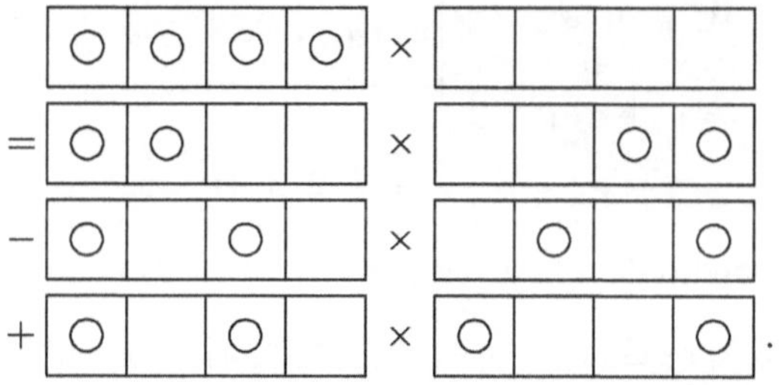

The purpose of this chapter is to understand the meaning of the above diagrammatic expression. To this end, it is necessary to learn some facts about pfaffians and determinants.

2.2 Pfaffians

Although the properties of determinants are well known, most people know little about pfaffians [30]. This section discusses the properties of pfaffians, which are more varied than those of a determinant. Determinantal identities, such as Plücker relations and Jacobi identities, are extended and unified as pfaffian identities, which are, in fact, simpler. There are many other interesting features of pfaffians which have been discovered (or rediscovered) through research into soliton equations. The author is certain that, as pfaffians become more widely known, their applications in other fields will be developed further.

A pfaffian may be obtained from a determinant in the following way. Let A be the determinant of an $m \times m$ antisymmetric matrix,

$$A = \det(a_{j,k}) \quad (1 \le j, k \le m), \tag{2.1}$$

where $a_{j,k} = -a_{k,j}$. If m is odd, then $A = 0$, and if m is even, then A is the square of a pfaffian. This pfaffian has order n, where $m = 2n$, thought of as a polynomial in matrix entries, and is denoted

$$(1, 2, \ldots, 2n).$$

For example, if $n = 1$,

$$\begin{vmatrix} 0 & a_{1,2} \\ -a_{1,2} & 0 \end{vmatrix} = a_{1,2}^2 \equiv (1, 2)^2, \tag{2.2}$$

and, if $n = 2$,

$$\begin{vmatrix} 0 & a_{1,2} & a_{1,3} & a_{1,4} \\ -a_{1,2} & 0 & a_{2,3} & a_{2,4} \\ -a_{1,3} & -a_{2,3} & 0 & a_{3,4} \\ -a_{1,4} & -a_{2,4} & -a_{3,4} & 0 \end{vmatrix} = (a_{1,2}a_{3,4} - a_{1,3}a_{2,4} + a_{1,4}a_{2,3})^2$$
$$\equiv (1, 2, 3, 4)^2. \tag{2.3}$$

Therefore, a second-order pfaffian given by $(1, 2, 3, 4)$ is expanded as

$$(1, 2, 3, 4) = (1, 2)(3, 4) - (1, 3)(2, 4) + (1, 4)(2, 3), \tag{2.4}$$

where $(j, k) = a_{j,k}$ for $j < k$. It should be noted that, from the antisymmetric property $a_{k,j} = -a_{j,k}$, we have

$$(k, j) = -(j, k). \tag{2.5}$$

In general, a pfaffian $(1, 2, \ldots, 2n)$ can be expanded as

$$\begin{aligned}(1, 2, \ldots, 2n) &= (1, 2)(3, 4, \ldots, 2n) - (1, 3)(2, 4, 5, \ldots, 2n) \\ &\quad + (1, 4)(2, 3, 5, \ldots, 2n) - \cdots \\ &\quad + (1, 2n)(2, 3, \ldots, 2n - 1) \\ &= \sum_{j=2}^{2n} (-1)^j (1, j)(2, 3, \ldots, \widehat{j}, \ldots, 2n),\end{aligned} \tag{2.6}$$

where $\widehat{j}$ means that index j is omitted. An alternative expansion is given by

$$\begin{aligned}(1, 2, \ldots, 2n) &= (2, 3, \ldots, 2n-1)(1, 2n) \\ &\quad - (1, 3, 4, \ldots, 2n-1)(2, 2n) \\ &\quad + (1, 2, 4, \ldots, 2n-1)(3, 2n) \\ &\quad - \cdots + (1, 2, \ldots, 2n-2)(2n-1, 2n) \\ &= \sum_{j=1}^{2n-1} (-1)^{j-1}(1, 2, \ldots, \widehat{j}, \ldots, 2n-1)(j, 2n). \end{aligned} \tag{2.7}$$

Repeating the above expansion, we arrive at the summation of products of first-order pfaffians [30]:

$$(1, 2, \ldots, 2n) = \sum_P{}' (-1)^P (i_1, i_2)(i_3, i_4)(i_5, i_6) \cdots (i_{2n-1}, i_{2n}). \tag{2.8}$$

These first-order pfaffians (i, j) are called the *entries* in the pfaffian. In the above equation, $\sum'$ means the sum over all possible combinations of pairs selected from $\{1, 2, \ldots, 2n\}$ that satisfy

$$\begin{gathered} i_1 < i_2, \;\; i_3 < i_4, \;\; i_5 < i_6, \ldots, \;\; i_{2n-1} < i_{2n}, \\ i_1 < i_3 < i_5 < \cdots < i_{2n-1}. \end{gathered} \tag{2.9}$$

The factor $(-1)^P$ takes the value $+1$ (-1) if the sequence $i_1, i_2, \ldots, i_{2n}$ is an even (odd) permutation of $1, 2, \ldots, 2n$.

Remarks

(1) Since the expansion formula (2.8) is hardly ever used in soliton theory, it is not necessary to memorize it exactly. However, it should be remarked that this gives another definition of a pfaffian.

(2) The expansion formula (2.8) has a similar structure to that of an nth-order determinant,

$$\begin{aligned}\det(a_{j,k})_{1\le j,k\le n} &= \begin{vmatrix} a_{1,1} & a_{1,2} & a_{1,3} & \cdots & a_{1,n} \\ a_{2,1} & a_{2,2} & a_{2,3} & \cdots & a_{2,n} \\ a_{3,1} & a_{3,2} & a_{3,3} & \cdots & a_{3,n} \\ \vdots & \vdots & \vdots & & \vdots \\ a_{n,1} & a_{n,2} & a_{n,3} & \cdots & a_{n,n} \end{vmatrix} \\ &= \sum_P{}' (-1)^P a_{i_1,1} a_{i_2,2} a_{i_3,3} \cdots a_{i_n,n}, \end{aligned} \tag{2.10}$$

where here the sum is over *all* permutations $i_1, i_2, i_3, \ldots, i_n$ of 1, 2, 3, …, n and $(-1)^P$ is as in (2.8).

(3) An nth-order pfaffian is expressed using $2n$ indices $(1, 2, \ldots, 2n)$. The number of independent entries (i, j) is equal to ${}_{2n}C_2$, that is, the number of ways to choose two objects from $2n$.

(4) If two indices among the $2n$ coincide, then the value of the pfaffian is equal to zero.

(5) We may also use other notation for the pfaffian $(1, 2, \ldots, 2n)$.

(i) In order to specify all the entries, we adopt the following notation, where we use a triangular array, containing the upper triangular entries of a matrix:

$$(1, 2, 3, 4, \ldots, 2n) = \left| \begin{matrix} a_{1,2} & a_{1,3} & a_{1,4} & \cdots & a_{1,2n} \\ & a_{2,3} & a_{2,4} & \cdots & a_{2,2n} \\ & & a_{3,4} & \cdots & a_{3,2n} \\ & & & \ddots & \vdots \\ & & & & a_{2n-1,2n} \end{matrix} \right| . \qquad (2.11)$$

(ii) In analogy with the notation for an nth-order determinant, we sometimes employ the notation

$$(1, 2, 3, 4, \ldots, 2n) = \mathrm{Pf}\,(a_{i,j})_{1 \le i < j \le 2n}.$$

(6) Johann Friedrich Pfaff (1765–1825) is famous for the Pfaff form of an ordinary differential equation. He was known to be a judge of the doctoral dissertation of the great mathematician Karl Friedrich Gauss. □

2.3 Exterior algebra

Using exterior algebra, which arises in connection with the vector (exterior) product satisfying $A \times B = -B \times A$, one can give a clearer definition of a determinant and of a pfaffian. First of all, let us introduce a one-form given by

$$\omega_i = \sum_{j=1}^{n} a_{i,j} x^j \quad (i = 1, 2, \ldots, 2n). \qquad (2.12)$$

The fundamental property is that the x^j's satisfy the antisymmetric commutation relations

$$x^i \wedge x^j = -x^j \wedge x^i, \quad x^i \wedge x^i = 0. \qquad (2.13)$$

Apart from the above relations, such an object obeys the normal rules of algebra.

Remark

We use the product symbol $\wedge$ instead of $\times$. □

The coefficients $a_{i,j}$ are arbitrary complex functions, and the determinant $\det(a_{i,j})_{1\le i,j\le n}$ is defined by means of the exterior product of n one-forms:

$$\omega_1 \wedge \omega_2 \wedge \omega_3 \wedge \cdots \wedge \omega_n \equiv \det(a_{i,j})_{1\le i,j\le n} x^1 \wedge x^2 \wedge \cdots \wedge x^n. \tag{2.14}$$

For example, if $n = 2$,

$$\begin{aligned}\omega_1 \wedge \omega_2 &= (a_{1,1}x^1 + a_{1,2}x^2) \wedge (a_{2,1}x^1 + a_{2,2}x^2)\\ &= (a_{1,1}a_{2,2} - a_{1,2}a_{2,1})x^1 \wedge x^2\\ &= \begin{vmatrix} a_{1,1} & a_{1,2} \\ a_{2,1} & a_{2,2}\end{vmatrix} x^1 \wedge x^2.\end{aligned} \tag{2.15}$$

Next, let Ω be a two-form given by

$$\Omega = \sum_{1\le i<j\le 2n} b_{i,j}\, x^i \wedge x^j, \quad b_{i,j} = -b_{j,i}. \tag{2.16}$$

A pfaffian with (i,j) entry $b_{i,j}$ is defined in terms of the n-tuple exterior product of Ω:

$$\underbrace{\Omega \wedge \Omega \wedge \cdots \wedge \Omega}_{n \text{ copies}} \equiv n!\,(1, 2, 3, \ldots, 2n)\, x^1 \wedge x^2 \wedge \cdots \wedge x^{2n}. \tag{2.17}$$

From the above definition, one obtains the expansion formula for the pfaffian. For example, in the case $n = 2$, we have

$$\begin{aligned}\Omega = {}&(1, 2)x^1 \wedge x^2 + (1, 3)x^1 \wedge x^3 + (1, 4)x^1 \wedge x^4 + (2, 3)x^2 \wedge x^3\\ &+ (2, 4)x^2 \wedge x^4 + (3, 4)x^3 \wedge x^4,\end{aligned} \tag{2.18}$$

and so

$$\begin{aligned}\Omega \wedge \Omega = {}&\Big[(1, 2)x^1 \wedge x^2 + (1, 3)x^1 \wedge x^3 + (1, 4)x^1 \wedge x^4\\ &+ (2, 3)x^2 \wedge x^3 + (2, 4)x^2 \wedge x^4 + (3, 4)x^3 \wedge x^4\Big]\\ &\wedge \Big[(1, 2)x^1 \wedge x^2 + (1, 3)x^1 \wedge x^3 + (1, 4)x^1 \wedge x^4\\ &+ (2, 3)x^2 \wedge x^3 + (2, 4)x^2 \wedge x^4 + (3, 4)x^3 \wedge x^4\Big]\end{aligned}$$

$$
\begin{aligned}
&= (1,2)(3,4)\, x^1 \wedge x^2 \wedge x^3 \wedge x^4 + (1,3)(2,4)\, x^1 \wedge x^3 \wedge x^2 \wedge x^4 \\
&\quad + (1,4)(2,3)\, x^1 \wedge x^4 \wedge x^2 \wedge x^3 + (2,3)(1,4)\, x^2 \wedge x^3 \wedge x^1 \wedge x^4 \\
&\quad + (2,4)(1,3)\, x^2 \wedge x^4 \wedge x^1 \wedge x^3 + (3,4)(1,2)\, x^3 \wedge x^4 \wedge x^1 \wedge x^2 \\
&= 2\,[(1,2)(3,4) - (1,3)(2,4) + (1,4)(2,3)]\, x^1 \wedge x^2 \wedge x^3 \wedge x^4.
\end{aligned}
\tag{2.19}
$$

On the other hand, definition (2.17) gives

$$\Omega \wedge \Omega \equiv 2\,(1,2,3,4)\, x^1 \wedge x^2 \wedge x^3 \wedge x^4, \tag{2.20}$$

and so, comparing (2.19) and (2.20), we obtain the expansion expression

$$(1,2,3,4) = (1,2)(3,4) - (1,3)(2,4) + (1,4)(2,3). \tag{2.21}$$

2.4 Expressions for general determinants and wronskians

We have given a definition of the pfaffian through the determinant of a $2n \times 2n$ antisymmetric matrix. Conversely, an nth-order determinant,

$$B \equiv \det(b_{j,k})_{1\le j,k\le n}, \tag{2.22}$$

is expressed as an nth-order pfaffian,

$$B = (1, 2, \ldots, n, n^*, \ldots, 2^*, 1^*), \tag{2.23}$$

where the pfaffian entries (j,k), (j^*,k^*), (j,k^*) are defined by

$$(j,k) = 0, \quad (j^*,k^*) = 0, \quad (j,k^*) = b_{j,k}. \tag{2.24}$$

Remark

The superscript * was originally used in connection with creation–annihilation operators [30] in quantum field theory. In this book, however, it is simply used to distinguish j and j^*. □

For example, if $n = 2$, we have

$$\begin{vmatrix} b_{1,1} & b_{1,2} \\ b_{2,1} & b_{2,2} \end{vmatrix} = (1, 2, 2^*, 1^*). \tag{2.25}$$

This is because

$$\text{RHS} = -(1,2^*)(2,1^*) + (1,1^*)(2,2^*) = b_{1,1}b_{2,2} - b_{1,2}b_{2,1} = \text{LHS}.$$

Next we consider a *wronskian* determinant, which often appears in the theory of linear ordinary differential equations. An nth-order wronskian $\mathrm{Wr}(f_1, f_2, \dots, f_n)$ is defined by

$$\mathrm{Wr}(f_1, f_2, \dots, f_n) \equiv \det\left(\frac{\partial^{j-1}}{\partial x^{j-1}} f_i\right)_{1\le i,j\le n}$$

$$= \begin{vmatrix} f_1^{(0)} & f_1^{(1)} & f_1^{(2)} & \cdots & f_1^{(n-1)} \\ f_2^{(0)} & f_2^{(1)} & f_2^{(2)} & \cdots & f_2^{(n-1)} \\ f_3^{(0)} & f_3^{(1)} & f_3^{(2)} & \cdots & f_3^{(n-1)} \\ \vdots & \vdots & \vdots & & \vdots \\ f_n^{(0)} & f_n^{(1)} & f_n^{(2)} & \cdots & f_n^{(n-1)} \end{vmatrix}, \tag{2.26}$$

where $f_i^{(j)}$ stands for the jth derivative of $f_i = f_i(x)$ with respect to x,

$$f_i^{(j)} \equiv \frac{\partial^j}{\partial x^j} f_i \quad (j = 0, 1, 2, \dots, n-1). \tag{2.27}$$

Let us introduce pfaffian entries

$$(d_j, i) \equiv f_i^{(j)}, \qquad (d_j, d_k) \equiv 0, \tag{2.28}$$

for $i = 1, 2, \dots, n$ and $j, k = 0, 1, 2, \dots, n-1$. Using these, an nth-order wronskian can be expressed as an nth-order pfaffian [31]:

$$\mathrm{Wr}(f_1, f_2, \dots, f_n) = (d_0, d_1, d_2, \dots, d_{n-1}, n, \dots, 3, 2, 1). \tag{2.29}$$

For example, in the case of $n = 2$, we have

$$\begin{aligned} \begin{vmatrix} f_1^{(0)} & f_1^{(1)} \\ f_2^{(0)} & f_2^{(1)} \end{vmatrix} &= (d_0, d_1, 2, 1) \\ &= -(d_0, 2)(d_1, 1) + (d_0, 1)(d_1, 2) \\ &= f_1^{(0)} f_2^{(1)} - f_2^{(0)} f_1^{(1)}. \end{aligned} \tag{2.30}$$

We have already seen that any nth-order determinant can be expressed as an nth-order pfaffian. It can also be shown that a $2n$th-order determinant can also be expressed as an nth-order pfaffian. Using the exterior algebra described in Section 2.3, take $2n$ one-forms

$$\omega_i = \sum_{j=1}^{2n} a_{i,j} x^j \quad (i = 1, 2, \dots, 2n). \tag{2.31}$$

The determinant $\det(a_{i,j})_{1\le i,j\le 2n}$ can be defined by

$$\omega_1 \wedge \omega_2 \wedge \cdots \wedge \omega_{2n} \equiv \det(a_{i,j})_{1\le i,j\le 2n}\, x^1 \wedge x^2 \wedge \cdots \wedge x^{2n}. \qquad (2.32)$$

Next, recall that if Ω is the two-form

$$\Omega = \sum_{1\le i<j\le 2n} (i, j)\, x^i \wedge x^j,$$

then the pfaffian $(1, 2, 3, \ldots, 2n)$ is defined by the n-fold exterior product of Ω,

$$\underbrace{\Omega \wedge \Omega \wedge \cdots \wedge \Omega}_{n \text{ copies}} \equiv n!\, (1, 2, 3, \ldots, 2n)\, x^1 \wedge x^2 \wedge \cdots \wedge x^{2n}. \qquad (2.33)$$

Now consider the particular two-form

$$\Omega \equiv \omega_1 \wedge \omega_2 + \omega_3 \wedge \omega_4 + \cdots + \omega_{2n-1} \wedge \omega_{2n}. \qquad (2.34)$$

For this Ω,

$$\underbrace{\Omega \wedge \Omega \wedge \cdots \wedge \Omega}_{n \text{ copies}} = n!\, \omega_1 \wedge \omega_2 \wedge \omega_3 \wedge \cdots \wedge \omega_{2n}. \qquad (2.35)$$

Let us consider ω_i's defined by

$$\omega_i = \sum_{j=1}^{2n} a_{i,j} x^j, \quad (i = 1, 2, \ldots, 2n). \qquad (2.36)$$

From the definition, we have

$$\omega_1 \wedge \omega_2 \wedge \omega_3 \wedge \cdots \wedge \omega_{2n} = \det(a_{i,j})_{1\le i,j\le 2n}\, x^1 \wedge x^2 \wedge \cdots \wedge x^{2n}. \qquad (2.37)$$

Substituting the above into (2.35), we obtain

$$\underbrace{\Omega \wedge \Omega \wedge \cdots \wedge \Omega}_{n \text{ copies}} = n!\, \det(a_{i,j})_{1\le i,j\le 2n}\, x^1 \wedge x^2 \wedge x^3 \wedge \cdots \wedge x^{2n}. \qquad (2.38)$$

On the other hand, substituting (2.36) into (2.34) gives

$$\Omega = \sum_{1\le i<j\le 2n} b_{i,j}\, x^i \wedge x^j, \qquad (2.39)$$

where $b_{i,j}$'s are given by

$$b_{i,j} \equiv \sum_{m=1}^{n} (a_{2m-1,i}\, a_{2m,j} - a_{2m-1,j}\, a_{2m,i}).$$

It follows from (2.39) that

$$\underbrace{\Omega \wedge \Omega \wedge \cdots \wedge \Omega}_{n \text{ copies}} = n!\, (1, 2, \ldots, 2n)\, x^1 \wedge x^2 \wedge \cdots \wedge x^{2n}, \tag{2.40}$$

where $(i, j) \equiv b_{i,j}$ for $i, j = 1, 2, \ldots, 2n$. Comparing (2.38) and (2.40), we deduce that

$$\det(a_{i,j})_{1 \le i,j \le 2n} = (1, 2, \ldots, 2n), \tag{2.41}$$

which expresses any $2n$th-order determinant as an nth-order pfaffian. For example, if $n = 1$, we have

$$\begin{vmatrix} a_{1,1} & a_{1,2} \\ a_{2,1} & a_{2,2} \end{vmatrix} \equiv (1, 2) = a_{1,1}a_{2,2} - a_{1,2}a_{2,1}. \tag{2.42}$$

An odd-order determinant has the pfaffian representation

$$\det(a_{i,j})_{1 \le i,j \le 2n+1} = (d_0, 1, 2, \ldots, 2n+1), \tag{2.43}$$

with its entries given by

$$(i, j) = \sum_{m=1}^{n} (a_{2m-1,i}\, a_{2m,j} - a_{2m-1,j}\, a_{2m,i}), \quad (d_0, i) = a_{2n+1,i}, \tag{2.44}$$

that is, a $(2n + 1)$th-order determinant may be expressed as an $(n + 1)$th-order pfaffian.

For example, if $n = 1$, we have

$$\begin{aligned} \begin{vmatrix} a_{1,1} & a_{1,2} & a_{1,3} \\ a_{2,1} & a_{2,2} & a_{2,3} \\ a_{3,1} & a_{3,2} & a_{3,3} \end{vmatrix} &= a_{3,1}(a_{1,2}a_{2,3} - a_{1,3}a_{2,2}) - a_{3,2}(a_{1,1}a_{2,3} - a_{1,3}a_{2,1}) \\ &\quad + a_{3,3}(a_{1,1}a_{2,2} - a_{1,2}a_{2,1}) \\ &= (d_0, 1)(2, 3) - (d_0, 2)(1, 3) + (d_0, 3)(1, 2) \\ &= (d_0, 1, 2, 3). \end{aligned} \tag{2.45}$$

From this we observe that the pfaffian expression for an odd-order determinant is obtained by expanding the determinant by the last row, considering it as a sum of even-order determinants and then making use of a pfaffian expression for these even-order determinants.

2.5 Laplace expansions of determinants and Plücker relations

2.5.1 Laplace expansions of determinants

An nth-order determinant $A_n = \det(a_{i,j})_{1\le i,j\le n}$ can be expressed as a sum of products of rth- and $(n-r)$th-order determinants.

We first choose r indices $i_1 < i_2 < \cdots < i_r$ from $1, 2, 3, \ldots, n$, and let the remaining ones be $i_{r+1} < i_{r+2} < \cdots < i_n$. Then we have

$$A_n = \sum (-1)^P \Delta \begin{pmatrix} i_1 & i_2 & \cdots & i_r \\ j_1 & j_2 & \cdots & j_r \end{pmatrix} \Delta \begin{pmatrix} i_{r+1} & i_{r+2} & \cdots & i_n \\ j_{r+1} & j_{r+2} & \cdots & j_n \end{pmatrix}, \tag{2.46}$$

where $P = i_1 + i_2 + \cdots + i_r + j_1 + j_2 + \cdots + j_r$. In the above,

$$\Delta \begin{pmatrix} i_1 & i_2 & \cdots & i_r \\ j_1 & j_2 & \cdots & j_r \end{pmatrix}$$

denotes the (rth-order) determinant of the matrix $\left(a_{i_p,j_q}\right)_{1\le p,q\le r}$, and

$$\Delta \begin{pmatrix} i_{r+1} & i_{r+2} & \cdots & i_n \\ j_{r+1} & j_{r+2} & \cdots & j_n \end{pmatrix}$$

denotes the ($(n-r)$th-order) determinant of the matrix $\left(a_{i_p,j_q}\right)_{r+1\le p,q\le n}$. The indices $j_1 < j_2 < \cdots < j_r$ are chosen from $1, 2, \ldots, n$, and the remaining ones are $j_{r+1} < j_{r+2} < \cdots < j_n$, and $\sum$ denotes the sum over all possible such choices, ${}_nC_r = n!/[r!(n-r)!]$ in number, of $j_1, j_2, \ldots, j_r$. This expansion formula is called the *Laplace expansion* of A_n.

Remark

The Japanese mathematician Yoshita Kurushima (*d.*1757) discovered the Laplace expansion of a determinant before Pierre-Simon Laplace (1749–1827). □

We illustrate the Laplace expansion (2.46) by means of an example. A fourth-order determinant can be expanded as products of second-order determinants as

$$\begin{aligned}
\begin{vmatrix} a_{1,1} & a_{1,2} & a_{1,3} & a_{1,4} \\ a_{2,1} & a_{2,2} & a_{2,3} & a_{2,4} \\ a_{3,1} & a_{3,2} & a_{3,3} & a_{3,4} \\ a_{4,1} & a_{4,2} & a_{4,3} & a_{4,4} \end{vmatrix} &= \begin{vmatrix} a_{1,1} & a_{1,2} \\ a_{2,1} & a_{2,2} \end{vmatrix} \begin{vmatrix} a_{3,3} & a_{3,4} \\ a_{4,3} & a_{4,4} \end{vmatrix} \\
&- \begin{vmatrix} a_{1,1} & a_{1,3} \\ a_{2,1} & a_{2,3} \end{vmatrix} \begin{vmatrix} a_{3,2} & a_{3,4} \\ a_{4,2} & a_{4,4} \end{vmatrix} + \begin{vmatrix} a_{1,1} & a_{1,4} \\ a_{2,1} & a_{2,4} \end{vmatrix} \begin{vmatrix} a_{3,2} & a_{3,3} \\ a_{4,2} & a_{4,3} \end{vmatrix} \\
&+ \begin{vmatrix} a_{1,2} & a_{1,3} \\ a_{2,2} & a_{2,3} \end{vmatrix} \begin{vmatrix} a_{3,1} & a_{3,4} \\ a_{4,1} & a_{4,4} \end{vmatrix} - \begin{vmatrix} a_{1,2} & a_{1,4} \\ a_{2,2} & a_{2,4} \end{vmatrix} \begin{vmatrix} a_{3,1} & a_{3,3} \\ a_{4,1} & a_{4,3} \end{vmatrix} \\
&+ \begin{vmatrix} a_{1,3} & a_{1,4} \\ a_{2,3} & a_{2,4} \end{vmatrix} \begin{vmatrix} a_{3,1} & a_{3,2} \\ a_{4,1} & a_{4,2} \end{vmatrix}. \qquad (2.47)
\end{aligned}$$

This is an example of a Laplace expansion.

Let us establish the expansion formula (2.47) using exterior algebra. Let ω_i for $i = 1, 2, 3, 4$ be one-forms,

$$\omega_i = \sum_{j=1}^{4} a_{i,j}\, x^j \quad (i = 1, 2, 3, 4). \qquad (2.48)$$

Then, by definition, we have

$$\omega_1 \wedge \omega_2 \wedge \omega_3 \wedge \omega_4 = \begin{vmatrix} a_{1,1} & a_{1,2} & a_{1,3} & a_{1,4} \\ a_{2,1} & a_{2,2} & a_{2,3} & a_{2,4} \\ a_{3,1} & a_{3,2} & a_{3,3} & a_{3,4} \\ a_{4,1} & a_{4,2} & a_{4,3} & a_{4,4} \end{vmatrix} x^1 \wedge x^2 \wedge x^3 \wedge x^4. \qquad (2.49)$$

On the other hand,

$$\begin{aligned}
\omega_1 \wedge \omega_2 &= (a_{1,1}x^1 + a_{1,2}x^2 + a_{1,3}x^3 + a_{1,4}x^4) \\
&\quad \wedge (a_{2,1}x^1 + a_{2,2}x^2 + a_{2,3}x^3 + a_{2,4}x^4) \\
&= \begin{vmatrix} a_{1,1} & a_{1,2} \\ a_{2,1} & a_{2,2} \end{vmatrix} x^1 \wedge x^2 + \begin{vmatrix} a_{1,1} & a_{1,3} \\ a_{2,1} & a_{2,3} \end{vmatrix} x^1 \wedge x^3 \\
&\quad + \begin{vmatrix} a_{1,1} & a_{1,4} \\ a_{2,1} & a_{2,4} \end{vmatrix} x^1 \wedge x^4 + \begin{vmatrix} a_{1,2} & a_{1,3} \\ a_{2,2} & a_{2,3} \end{vmatrix} x^2 \wedge x^3 \\
&\quad + \begin{vmatrix} a_{1,2} & a_{1,4} \\ a_{2,2} & a_{2,4} \end{vmatrix} x^2 \wedge x^4 + \begin{vmatrix} a_{1,3} & a_{1,4} \\ a_{2,3} & a_{2,4} \end{vmatrix} x^3 \wedge x^4
\end{aligned}$$

and

$$\begin{aligned}\omega_3 \wedge \omega_4 &= (a_{3,1}x^1 + a_{3,2}x^2 + a_{3,3}x^3 + a_{3,4}x^4) \\ &\quad \wedge (a_{4,1}x^1 + a_{4,2}x^2 + a_{4,3}x^3 + a_{4,4}x^4) \\ &= \begin{vmatrix} a_{3,1} & a_{3,2} \\ a_{4,1} & a_{4,2} \end{vmatrix} x^1 \wedge x^2 + \begin{vmatrix} a_{3,1} & a_{3,3} \\ a_{4,1} & a_{4,3} \end{vmatrix} x^1 \wedge x^3 \\ &\quad + \begin{vmatrix} a_{3,1} & a_{3,4} \\ a_{4,1} & a_{4,4} \end{vmatrix} x^1 \wedge x^4 + \begin{vmatrix} a_{3,2} & a_{3,3} \\ a_{4,2} & a_{4,3} \end{vmatrix} x^2 \wedge x^3 \\ &\quad + \begin{vmatrix} a_{3,2} & a_{3,4} \\ a_{4,2} & a_{4,4} \end{vmatrix} x^2 \wedge x^4 + \begin{vmatrix} a_{3,3} & a_{3,4} \\ a_{4,3} & a_{4,4} \end{vmatrix} x^3 \wedge x^4.\end{aligned}$$

Substituting the above into

$$\omega_1 \wedge \omega_2 \wedge \omega_3 \wedge \omega_4 = (\omega_1 \wedge \omega_2) \wedge (\omega_3 \wedge \omega_4)$$

gives

$$\begin{aligned}&\omega_1 \wedge \omega_2 \wedge \omega_3 \wedge \omega_4 \\ &\quad = \left\{ \begin{vmatrix} a_{1,1} & a_{1,2} \\ a_{2,1} & a_{2,2} \end{vmatrix} \begin{vmatrix} a_{3,3} & a_{3,4} \\ a_{4,3} & a_{4,4} \end{vmatrix} - \begin{vmatrix} a_{1,1} & a_{1,3} \\ a_{2,1} & a_{2,3} \end{vmatrix} \begin{vmatrix} a_{3,2} & a_{3,4} \\ a_{4,2} & a_{4,4} \end{vmatrix} \right. \\ &\qquad + \begin{vmatrix} a_{1,1} & a_{1,4} \\ a_{2,1} & a_{2,4} \end{vmatrix} \begin{vmatrix} a_{3,2} & a_{3,3} \\ a_{4,2} & a_{4,3} \end{vmatrix} + \begin{vmatrix} a_{1,2} & a_{1,3} \\ a_{2,2} & a_{2,3} \end{vmatrix} \begin{vmatrix} a_{3,1} & a_{3,4} \\ a_{4,1} & a_{4,4} \end{vmatrix} \\ &\qquad \left. - \begin{vmatrix} a_{1,2} & a_{1,4} \\ a_{2,2} & a_{2,4} \end{vmatrix} \begin{vmatrix} a_{3,1} & a_{3,3} \\ a_{4,1} & a_{4,3} \end{vmatrix} + \begin{vmatrix} a_{1,3} & a_{1,4} \\ a_{2,3} & a_{2,4} \end{vmatrix} \begin{vmatrix} a_{3,1} & a_{3,2} \\ a_{4,1} & a_{4,2} \end{vmatrix} \right\} \\ &\qquad \times x^1 \wedge x^2 \wedge x^3 \wedge x^4.\end{aligned}$$

From (2.49), the terms inside the braces $\{\cdots\}$ are equal to the fourth-order determinant, which completes the proof of the Laplace expansion formula (2.47).

Finally, let us prove the general result. Taking the product of n one-forms,

$$\omega_i = \sum_{j=1}^{n} a_{i,j} x^j \quad (i = 1, 2, \ldots, n), \tag{2.50}$$

we generate an nth-order determinant,

$$\omega_1 \wedge \omega_2 \wedge \omega_3 \wedge \cdots \wedge \omega_n = \det(a_{i,j})_{1 \le i,j \le n} x^1 \wedge x^2 \wedge \cdots \wedge x^n. \tag{2.51}$$

The left-hand side may be written as the product

$$(\omega_1 \wedge \omega_2 \wedge \cdots \wedge \omega_r) \wedge (\omega_{r+1} \wedge \omega_{r+2} \wedge \cdots \wedge \omega_n),$$

which may be written as a sum of products of rth- and $(n-r)$th-order determinants. This establishes the formula (2.46) for the Laplace expansion.

2.5.2 Plücker relations

The following sum of products of second-order determinants,

$$\begin{vmatrix} a_0 & a_1 \\ b_0 & b_1 \end{vmatrix} \begin{vmatrix} a_2 & a_3 \\ b_2 & b_3 \end{vmatrix} - \begin{vmatrix} a_0 & a_2 \\ b_0 & b_2 \end{vmatrix} \begin{vmatrix} a_1 & a_3 \\ b_1 & b_3 \end{vmatrix} + \begin{vmatrix} a_0 & a_3 \\ b_0 & b_3 \end{vmatrix} \begin{vmatrix} a_1 & a_2 \\ b_1 & b_2 \end{vmatrix} = 0, \qquad (2.52)$$

is satisfied identically. This may be proved directly by expanding each determinant. There is, however, another method of proof. Consider the fourth-order determinant,

$$\begin{vmatrix} a_0 & a_1 & a_2 & a_3 \\ b_0 & b_1 & b_2 & b_3 \\ 0 & a_1 & a_2 & a_3 \\ 0 & b_1 & b_2 & b_3 \end{vmatrix} = \begin{vmatrix} a_0 & 0 & 0 & 0 \\ b_0 & 0 & 0 & 0 \\ 0 & a_1 & a_2 & a_3 \\ 0 & b_1 & b_2 & b_3 \end{vmatrix}. \qquad (2.53)$$

By means of the Laplace expansion theorem, the determinants on each side of this equation may be expanded to give

$$\begin{vmatrix} a_0 & a_1 \\ b_0 & b_1 \end{vmatrix} \begin{vmatrix} a_2 & a_3 \\ b_2 & b_3 \end{vmatrix} - \begin{vmatrix} a_0 & a_2 \\ b_0 & b_2 \end{vmatrix} \begin{vmatrix} a_1 & a_3 \\ b_1 & b_3 \end{vmatrix} + \begin{vmatrix} a_0 & a_3 \\ b_0 & b_3 \end{vmatrix} \begin{vmatrix} a_1 & a_2 \\ b_1 & b_2 \end{vmatrix} = 0, \qquad (2.54)$$

as required. Note that (2.54) can be expressed entirely in terms of the column vectors $\mathbf{c}_i = (a_i, b_i)^t$ as

$$|\mathbf{c}_0 \;\; \mathbf{c}_1|\,|\mathbf{c}_2 \;\; \mathbf{c}_3| - |\mathbf{c}_0 \;\; \mathbf{c}_2|\,|\mathbf{c}_1 \;\; \mathbf{c}_3| + |\mathbf{c}_0 \;\; \mathbf{c}_3|\,|\mathbf{c}_1 \;\; \mathbf{c}_2| = 0. \qquad (2.55)$$

This is the simplest case of a *Plücker relation.*

In fact, only the indices are important in (2.55), and so we may also express it as

$$(0\ 1)(2\ 3) - (0\ 2)(1\ 3) + (0\ 3)(1\ 2) = 0. \qquad (2.56)$$

The above expression may also be written in diagrammatic form using Sato's Maya diagrams

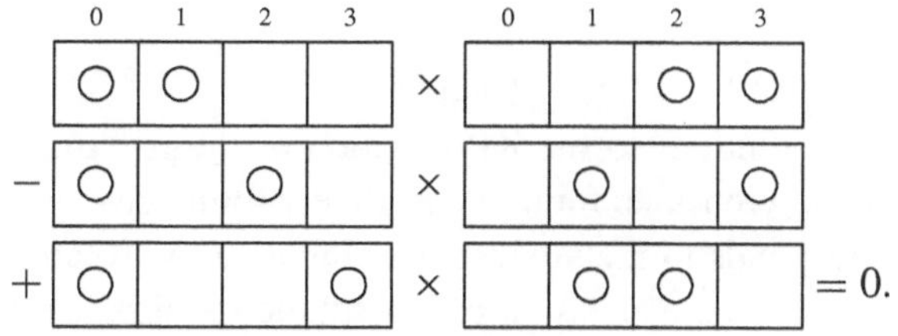

In the same way, the identity

$$\begin{vmatrix} f & a_0 & a_1 \\ g & b_0 & b_1 \\ h & c_0 & c_1 \end{vmatrix}\begin{vmatrix} f & a_2 & a_3 \\ g & b_2 & b_3 \\ h & c_2 & c_3 \end{vmatrix} - \begin{vmatrix} f & a_0 & a_2 \\ g & b_0 & b_2 \\ h & c_0 & c_2 \end{vmatrix}\begin{vmatrix} f & a_1 & a_3 \\ g & b_1 & b_3 \\ h & c_1 & c_3 \end{vmatrix}$$

$$+ \begin{vmatrix} f & a_0 & a_3 \\ g & b_0 & b_3 \\ h & c_0 & c_3 \end{vmatrix}\begin{vmatrix} f & a_1 & a_2 \\ g & b_1 & b_2 \\ h & c_1 & c_2 \end{vmatrix} = 0, \tag{2.57}$$

for a sum of products of third-order determinants follows from the Laplace expansion of the sixth-order determinant

$$\begin{vmatrix} f & a_0 & a_1 & 0 & a_2 & a_3 \\ g & b_0 & b_1 & 0 & b_2 & b_3 \\ h & c_0 & c_1 & 0 & c_2 & c_3 \\ 0 & 0 & a_1 & f & a_2 & a_3 \\ 0 & 0 & b_1 & g & b_2 & b_3 \\ 0 & 0 & c_1 & h & c_2 & c_3 \end{vmatrix}, \tag{2.58}$$

which is zero. Equation (2.57) may be expressed simply as

$$(f\ 0\ 1)(f\ 2\ 3) - (f\ 0\ 2)(f\ 1\ 3) + (f\ 0\ 3)(f\ 1\ 2) = 0, \tag{2.59}$$

and can be extended to

$$\begin{aligned}(f_1\ f_2\ \cdots\ f_N\ 0\ 1)(f_1\ f_2\ \cdots\ f_N\ 2\ 3) - (f_1\ f_2\ \cdots\ f_N\ 0\ 2)\\ \times (f_1\ f_2\ \cdots\ f_N\ 1\ 3) + (f_1\ f_2\ \cdots\ f_N\ 0\ 3)(f_1\ f_2\ \cdots\ f_N\ 1\ 2) = 0,\end{aligned} \tag{2.60}$$

which can be expressed in terms of Maya diagrams as

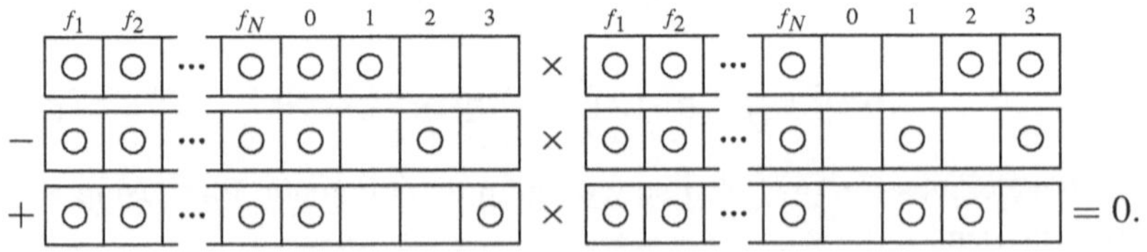

Remark

If one expresses solutions to the KP equation or the two-dimensional Toda lattice equation in wronskian form, their bilinear forms give this identity. The number N corresponds to the number of solitons in the solution. One of the merits in the wronskian expression of the solution is that the verification of the N-soliton solution can be performed in the same way as the two-soliton solution. □

Here let us discuss the relationship between Sato's Maya diagrams and Young diagrams. A Maya diagram describes a state in which fermions are distributed in a one-dimensional array of cells. Only one fermion can occupy each cell. We suppose that cells sufficiently far to the left are occupied with particles and that cells sufficiently far to the right are empty. This corresponds to a state in which fermions have been excited from a vacuum state. We distinguish occupied and empty cells using the symbols ○ and ×, respectively. Even though we have already denoted an empty cell by leaving it blank, we will use the notation × in order to clarify the correspondence with a Young diagram. An arbitrary Maya diagram,

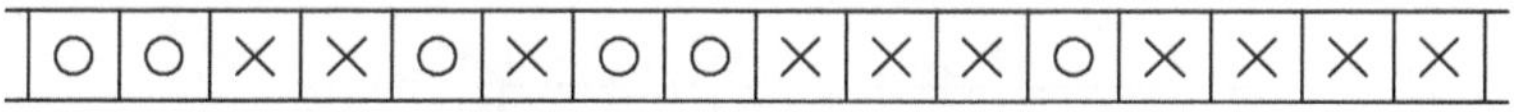

corresponds to the Young diagram [32],

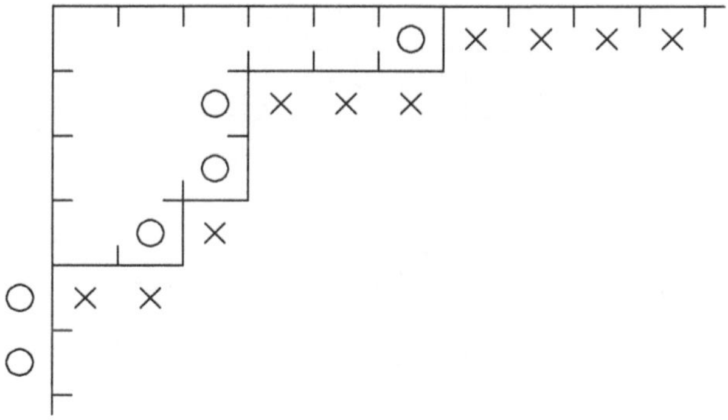

That is, ○ and × correspond to | and — in the Young diagram, respectively. Therefore, if one replaces each element in the Plücker relation,

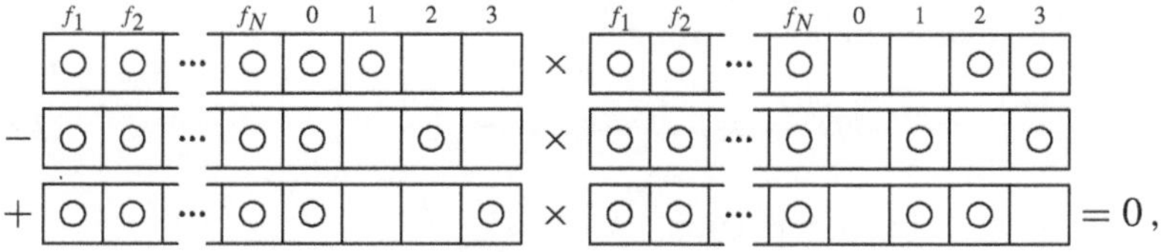

with

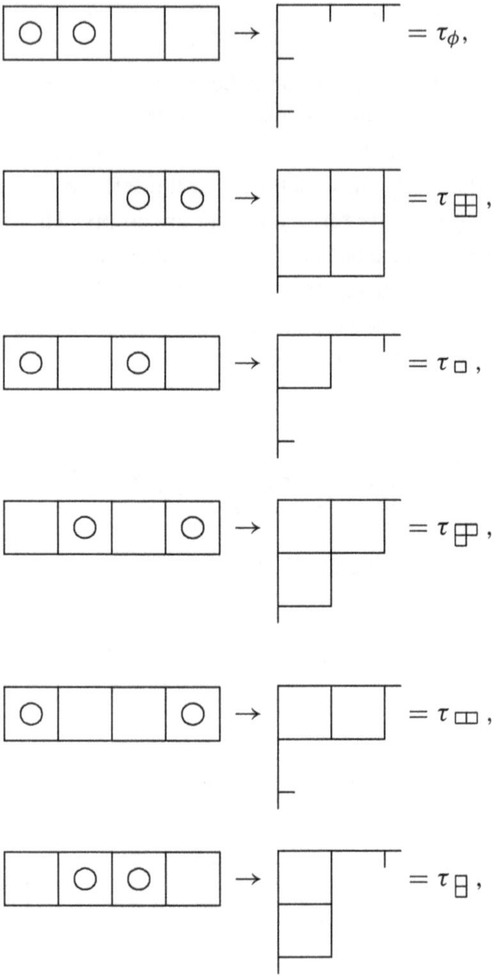

we have

$$\tau_{\phi}\tau_{(2,2)} - \tau_{(1)}\tau_{(2,1)} + \tau_{(2)}\tau_{(1,1)} = 0,$$

where ϕ stands for the vacuum state. It was Sato [12, 13] who discovered first that the KP equation in bilinear form,

$$(D_1^4 - 4D_1D_3 + 3D_2^2)\tau \cdot \tau = 0, \tag{2.61}$$

was nothing but a Plücker relation.

2.6 Jacobi identities for determinants

2.6.1 Cofactors

Consider a matrix $A = (a_{i,j})_{1 \leq i,j \leq n}$ whose determinant is D. The cofactor $\Delta_{i,j}$ with respect to $a_{i,j}$ is the determinant of the matrix obtained by eliminating the ith row and jth column from A, multiplied by $(-1)^{i+j}$. Then the following expansion formulae hold:

$$\begin{aligned} D &= \sum_{i=1}^{n} a_{i,j} \Delta_{i,j} \quad (j = 1, 2, \ldots, n) \\ &= \sum_{j=1}^{n} a_{i,j} \Delta_{i,j} \quad (i = 1, 2, \ldots, n). \end{aligned} \tag{2.62}$$

These are special cases of the orthogonality relations

$$\sum_{i=1}^{n} a_{i,j} \Delta_{i,k} = \delta_{j,k} D, \tag{2.63a}$$

$$\sum_{j=1}^{n} a_{i,j} \Delta_{k,j} = \delta_{i,k} D, \tag{2.63b}$$

where $\delta_{i,j}$ is the Kronecker δ symbol, equal to 1 if $i = j$ and 0 otherwise.

Similarly, let us introduce the cofactor $\Gamma(i, j)$ with respect to (i, j) in the nth-order pfaffian $(1, 2, 3, \ldots, 2n)$ defined by

$$\begin{aligned} &\Gamma(i, j) = (-1)^{i+j} (1, 2, \ldots, \widehat{i}, \ldots, \widehat{j}, \ldots, 2n) \quad (i < j), \\ &\Gamma(j, i) = -\Gamma(i, j), \quad \Gamma(i, i) = 0. \end{aligned} \tag{2.64}$$

In the case $n = 1$, the cofactor of $(1, 2)$, $\Gamma(1, 2)$, is defined to be unity.

The formula for expanding a pfaffian by its cofactors is

$$(1, 2, 3, \ldots, 2n) = \sum_{j=1}^{2n} (i, j) \, \Gamma(i, j) \quad (i = 1, 2, \ldots, 2n). \tag{2.65}$$

As for a determinant, the cofactors of a pfaffian satisfy orthogonality relations,

$$\sum_{i=1}^{2n} (i, j) \, \Gamma(i, k) = \delta_{j,k} P, \tag{2.66a}$$

$$\sum_{j=1}^{2n} (i, j) \, \Gamma(k, j) = \delta_{i,k} P, \tag{2.66b}$$

where $P \equiv (1, 2, 3, \ldots, 2n)$. For example, in the case $i = 1$, $k = 2$, the pfaffian orthogonality relation (2.66b) is

$$\sum_{j=3}^{2n} (1, j)\, \Gamma(2, j) = 0.$$

This is simply an expansion formula of the vanishing pfaffian $(1, 1, 3, 4, \ldots, 2n)$.

Now let us consider the case where n, the order of a determinant D, is even and each matrix element $a_{i,j}$ equals the pfaffian entry (i, j). The orthogonality relations (2.63) may be rewritten as

$$\sum_{i=1}^{n} a_{i,j} \Delta_{i,k} = \delta_{j,k} P^2,$$
$$\sum_{j=1}^{n} a_{i,j} \Delta_{k,j} = \delta_{i,k} P^2. \tag{2.67}$$

Comparing the above formulae with the pfaffian orthogonality relations (2.66), we obtain

$$\Delta_{i,k} = \Gamma(i, k) P. \tag{2.68}$$

Since the pfaffian cofactors $\Gamma(i, j)$ are, by definition, antisymmetric, we can consider the mth-order pfaffian with entries $\Gamma(i, j)$,

$$\Gamma(1, 2, 3, 4, \ldots, 2m) = \begin{vmatrix} \Gamma(1,2) & \Gamma(1,3) & \Gamma(1,4) & \cdots & \Gamma(1,2m) \\ & \Gamma(2,3) & \Gamma(2,4) & \cdots & \Gamma(2,2m) \\ & & \Gamma(3,4) & \cdots & \Gamma(3,2m) \\ & & & \ddots & \vdots \\ & & & & \Gamma(2m-1,2m) \end{vmatrix}. \tag{2.69}$$

Remark

The expansion formula of the above pfaffian is given by

$$\Gamma(1, 2, 3, \ldots, 2m) = \sum_{j=1}^{2m} (-1)^j \Gamma(1, j) \Gamma(2, 3, \ldots, \widehat{j}, \ldots, 2m). \tag{2.70}$$

In fact, we can also obtain pfaffian identities by reinterpreting this expansion formula. □

2.6.2 Jacobi identities for determinants [33]

The $(N-1)$th-order determinant obtained by eliminating the jth row and the kth column from an Nth-order determinant $D = \det(a_{i,j})_{1\le i,j\le N}$ is called the (j,k)*th minor* of D, which is denoted by $D\begin{bmatrix} j \\ k \end{bmatrix}$. As defined above, the cofactor Δ_{jk} equals $D\begin{bmatrix} j \\ k \end{bmatrix}$ multiplied by a signature $(-1)^{j+k}$. That is,

$$\Delta_{j,k} \equiv (-1)^{j+k} D\begin{bmatrix} j \\ k \end{bmatrix}, \tag{2.71}$$

where

$$D\begin{bmatrix} j \\ k \end{bmatrix} \equiv \left| \begin{array}{ccccc|c|ccc} a_{1,1} & a_{1,2} & a_{1,3} & \cdots & a_{1,k-1} & k & a_{1,k+1} & \cdots & a_{1,N} \\ a_{2,1} & a_{2,2} & a_{2,3} & \cdots & a_{2,k-1} & & a_{2,k+1} & \cdots & a_{2,N} \\ \vdots & \vdots & \vdots & & \vdots & & \vdots & & \vdots \\ a_{j-1,1} & a_{j-1,2} & a_{j-1,3} & \cdots & a_{j-1,k-1} & & a_{j-1,k+1} & \cdots & a_{j-1,N} \\ \hline j & & & & & & & & \\ \hline a_{j+1,1} & a_{j+1,2} & a_{j+1,3} & \cdots & a_{j+1,k-1} & & a_{j+1,k+1} & \cdots & a_{j+1,N} \\ \vdots & \vdots & \vdots & & \vdots & & \vdots & & \vdots \\ a_{N,1} & a_{N,2} & a_{N,3} & \cdots & a_{N,k-1} & & a_{N,k+1} & \cdots & a_{N,N} \end{array} \right|, \tag{2.72}$$

where the shading indicates the row and column that have been removed. In the same way, we denote the $(N-2)$nd-order determinant obtained by eliminating the jth and kth rows and the lth and mth columns from the determinant D as $D\begin{bmatrix} j & k \\ l & m \end{bmatrix}$. The formula

$$D\begin{bmatrix} i \\ i \end{bmatrix} D\begin{bmatrix} j \\ j \end{bmatrix} - D\begin{bmatrix} i \\ j \end{bmatrix} D\begin{bmatrix} j \\ i \end{bmatrix} = D\begin{bmatrix} i & j \\ i & j \end{bmatrix} D, \tag{2.73}$$

which is called the *Jacobi identity* [33], will be proved below.

First, let us write out (2.73) explicitly in the case $N = 3$:

$$D\begin{bmatrix}2\\2\end{bmatrix} D\begin{bmatrix}3\\3\end{bmatrix} - D\begin{bmatrix}2\\3\end{bmatrix} D\begin{bmatrix}3\\2\end{bmatrix} = D\begin{bmatrix}2 & 3\\2 & 3\end{bmatrix} D. \tag{2.74}$$

The determinant D is given by

$$D = \begin{vmatrix} a_{1,1} & a_{1,2} & a_{1,3} \\ a_{2,1} & a_{2,2} & a_{2,3} \\ a_{3,1} & a_{3,2} & a_{3,3} \end{vmatrix}, \tag{2.75}$$

and the minor $D\begin{bmatrix}2\\2\end{bmatrix}$, obtained by eliminating the second row and the second column from D, is

$$D\begin{bmatrix}2\\2\end{bmatrix} = \begin{vmatrix} a_{1,1} & a_{1,3} \\ a_{3,1} & a_{3,3} \end{vmatrix}. \tag{2.76}$$

Similarly, we have

$$\begin{aligned} D\begin{bmatrix}3\\3\end{bmatrix} &= \begin{vmatrix} a_{1,1} & a_{1,2} \\ a_{2,1} & a_{2,2} \end{vmatrix}, \quad D\begin{bmatrix}2\\3\end{bmatrix} = \begin{vmatrix} a_{1,1} & a_{1,2} \\ a_{3,1} & a_{3,2} \end{vmatrix}, \\ D\begin{bmatrix}3\\2\end{bmatrix} &= \begin{vmatrix} a_{1,1} & a_{1,3} \\ a_{2,1} & a_{2,3} \end{vmatrix}, \quad D\begin{bmatrix}2\ 3\\2\ 3\end{bmatrix} = a_{1,1}. \end{aligned} \tag{2.77}$$

Hence, the Jacobi identity is rewritten as

$$\begin{aligned} &\begin{vmatrix} a_{1,1} & a_{1,3} \\ a_{3,1} & a_{3,3} \end{vmatrix} \begin{vmatrix} a_{1,1} & a_{1,2} \\ a_{2,1} & a_{2,2} \end{vmatrix} - \begin{vmatrix} a_{1,1} & a_{1,2} \\ a_{3,1} & a_{3,2} \end{vmatrix} \begin{vmatrix} a_{1,1} & a_{1,3} \\ a_{2,1} & a_{2,3} \end{vmatrix} \\ &\quad = a_{1,1} \begin{vmatrix} a_{1,1} & a_{1,2} & a_{1,3} \\ a_{2,1} & a_{2,2} & a_{2,3} \\ a_{3,1} & a_{3,2} & a_{3,3} \end{vmatrix}. \end{aligned} \tag{2.78}$$

The above formula can be verified by direct calculation. Even though we can also make the same verification in the cases $N = 3, 4, 5$ by using computer algebra, it is still necessary to prove the formula for all N. The following proof is known [33].

We consider the product of D and an Nth-order determinant of cofactors, which we denote by Δ_{22}:

$$D\Delta_{22} = \begin{vmatrix} A_{11} & A_{12} \\ A_{21} & A_{22} \end{vmatrix} \begin{vmatrix} E & \Delta_{12} \\ 0 & \Delta_{22} \end{vmatrix}$$

$$\equiv \begin{vmatrix} a_{1,1} & \cdots & a_{1,r} & a_{1,r+1} & \cdots & a_{1,n} \\ a_{2,1} & \cdots & a_{2,r} & a_{2,r+1} & \cdots & a_{2,n} \\ \vdots & & \vdots & \vdots & & \vdots \\ a_{r,1} & \cdots & a_{r,r} & a_{r,r+1} & \cdots & a_{r,n} \\ a_{r+1,1} & \cdots & a_{r+1,r} & a_{r+1,r+1} & \cdots & a_{r+1,n} \\ a_{r+2,1} & \cdots & a_{r+2,r} & a_{r+2,r+1} & \cdots & a_{r+2,n} \\ \vdots & & \vdots & \vdots & & \vdots \\ a_{n,1} & \cdots & a_{nr} & a_{n,r+1} & \cdots & a_{n,n} \end{vmatrix}$$

$$\times \begin{vmatrix} 1 & 0 & \cdots & 0 & \Delta_{r+1,1} & \cdots & \Delta_{n,1} \\ 0 & 1 & \ddots & \vdots & \Delta_{r+1,2} & \cdots & \Delta_{n,2} \\ \vdots & \ddots & \ddots & 0 & \vdots & & \vdots \\ 0 & \cdots & 0 & 1 & \Delta_{r+1,r} & \cdots & \Delta_{n,r} \\ 0 & 0 & \cdots & 0 & \Delta_{r+1,r+1} & \cdots & \Delta_{n,r+1} \\ 0 & 0 & \cdots & 0 & \Delta_{r+1,r+2} & \cdots & \Delta_{n,r+2} \\ \vdots & \vdots & & \vdots & \vdots & \cdots & \vdots \\ 0 & 0 & \cdots & 0 & \Delta_{r+1,n} & \cdots & \Delta_{n,n} \end{vmatrix}, \tag{2.79}$$

where E is the $r \times r$ identity matrix, 0 is the $(n-r) \times r$ zero matrix, and

$$A_{11} \equiv \begin{pmatrix} a_{1,1} & \cdots & a_{1,r} \\ a_{2,1} & \cdots & a_{2,r} \\ \vdots & & \vdots \\ a_{r,1} & \cdots & a_{r,r} \end{pmatrix}, \qquad A_{12} \equiv \begin{pmatrix} a_{1,r+1} & \cdots & a_{1,n} \\ a_{2,r+1} & \cdots & a_{2,n} \\ \vdots & & \vdots \\ a_{r,r+1} & \cdots & a_{r,n} \end{pmatrix},$$

$$A_{21} \equiv \begin{pmatrix} a_{r+1,1} & \cdots & a_{r+1,r} \\ a_{r+2,1} & \cdots & a_{r+2,r} \\ \vdots & & \vdots \\ a_{n,1} & \cdots & a_{n,r} \end{pmatrix}, \qquad A_{22} \equiv \begin{pmatrix} a_{r+1,r+1} & \cdots & a_{r+1,n} \\ a_{r+2,r+1} & \cdots & a_{r+2,n} \\ \vdots & & \vdots \\ a_{n,r+1} & \cdots & a_{n,n} \end{pmatrix},$$

$$\Delta_{12} \equiv \begin{pmatrix} \Delta_{r+1,1} & \cdots & \Delta_{n,1} \\ \Delta_{r+1,2} & \cdots & \Delta_{n,2} \\ \vdots & & \vdots \\ \Delta_{r+1,r} & \cdots & \Delta_{n,r} \end{pmatrix}, \qquad \Delta_{22} \equiv \begin{pmatrix} \Delta_{r+1,r+1} & \cdots & \Delta_{n,r+1} \\ \Delta_{r+1,r+2} & \cdots & \Delta_{n,r+2} \\ \vdots & & \vdots \\ \Delta_{r+1,n} & \cdots & \Delta_{n,n} \end{pmatrix}.$$

Because of the orthogonality of the matrix entries $a_{i,j}$ and the cofactors $\Delta_{i,j}$, we have

$$\begin{vmatrix} A_{11} & A_{12} \\ A_{21} & A_{22} \end{vmatrix} \begin{vmatrix} E & \Delta_{12} \\ 0 & \Delta_{22} \end{vmatrix} = \begin{vmatrix} A_{11} & \mathbf{0} \\ A_{21} & D\mathbf{E} \end{vmatrix}, \tag{2.80}$$

where now $\mathbf{0}$ is the $r \times (n-r)$ zero matrix, $\mathbf{E}$ is the $(n-r) \times (n-r)$ identity matrix. The right-hand side of (2.80) written explicitly is

$$\begin{vmatrix} a_{1,1} & \cdots & a_{1,r} & 0 & \cdots & 0 \\ a_{2,1} & \cdots & a_{2,r} & 0 & \cdots & 0 \\ \vdots & & \vdots & \vdots & & \vdots \\ a_{r,1} & \cdots & a_{r,r} & 0 & \cdots & 0 \\ a_{r+1,1} & \cdots & a_{r+1,r} & D & \cdots & 0 \\ \vdots & & \vdots & \vdots & \ddots & \vdots \\ a_{n,1} & \cdots & a_{n,r} & 0 & \cdots & D \end{vmatrix}. \tag{2.80$'$}$$

Hence, (2.80) is equivalent to

$$D \begin{vmatrix} \Delta_{r+1,r+1} & \cdots & \Delta_{n,r+1} \\ \Delta_{r+1,r+2} & \cdots & \Delta_{n,r+2} \\ \vdots & & \vdots \\ \Delta_{r+1,n} & \cdots & \Delta_{nn} \end{vmatrix} = \begin{vmatrix} a_{1,1} & \cdots & a_{1,r} \\ a_{2,1} & \cdots & a_{2,r} \\ \vdots & & \vdots \\ a_{r,1} & \cdots & a_{r,r} \end{vmatrix} D^{n-r},$$

or, equivalently,

$$\begin{vmatrix} \Delta_{r+1,r+1} & \cdots & \Delta_{n,r+1} \\ \Delta_{r+1,r+2} & \cdots & \Delta_{n,r+2} \\ \vdots & & \vdots \\ \Delta_{r+1,n} & \cdots & \Delta_{n,n} \end{vmatrix} = \begin{vmatrix} a_{1,1} & \cdots & a_{1,r} \\ a_{2,1} & \cdots & a_{2,r} \\ \vdots & & \vdots \\ a_{r,1} & \cdots & a_{r,r} \end{vmatrix} D^{n-r-1}. \tag{2.81}$$

Putting $r = n - 2$, this gives

$$\Delta_{n-1,n-1}\Delta_{n,n} - \Delta_{n-1,n}\Delta_{n,n-1} = \begin{vmatrix} a_{1,1} & \cdots & a_{1,n-2} \\ a_{2,1} & \cdots & a_{2,n-2} \\ \vdots & & \vdots \\ a_{n-2,1} & \cdots & a_{n-2,n-2} \end{vmatrix} D, \tag{2.82}$$

which is nothing but the Jacobi identity,

$$D\begin{bmatrix} n-1 \\ n-1 \end{bmatrix} D\begin{bmatrix} n \\ n \end{bmatrix} - D\begin{bmatrix} n-1 \\ n \end{bmatrix} D\begin{bmatrix} n \\ n-1 \end{bmatrix} = D\begin{bmatrix} n-1 & n \\ n-1 & n \end{bmatrix} D. \qquad (2.83)$$

Snake's legs

Looking at the proof of the general case, it is difficult to understand the Jacobi identity immediately. The author came to understand the result by checking the formulae using computer algebra, looking for an alternative proof and applying it to actual problems. □

Remarks

(1) By employing the pfaffian expression for determinants given in Section 2.4, the terms in the above Jacobi identity are expressed as

$$D\begin{bmatrix} n-1 \\ n-1 \end{bmatrix} = (1, 2, \cdots, n-2, n, n^*, n-2^*, \ldots, 2^*, 1^*),$$

$$D\begin{bmatrix} n \\ n \end{bmatrix} = (1, 2, \ldots, n-2, n-1, n-1^*, n-2^*, \ldots, 2^*, 1^*),$$

$$D\begin{bmatrix} n-1 \\ n \end{bmatrix} = (1, 2, \ldots, n-2, n, n-1^*, n-2^*, \ldots, 2^*, 1^*),$$

$$D\begin{bmatrix} n \\ n-1 \end{bmatrix} = (1, 2, \ldots, n-2, n-1, n^*, n-2^*, \ldots, 2^*, 1^*),$$

$$D\begin{bmatrix} n-1 & n \\ n-1 & n \end{bmatrix} = (1, 2, \ldots, n-2, n-2^*, \ldots, 2^*, 1^*),$$

$$D = (1, 2, \ldots, n-2, n-1, n, n^*, n-1^*, n-2^*, \ldots, 2^*, 1^*),$$

where

$$(j, k) = (j^*, k^*) = 0, \quad (j, k^*) = a_{j,k}.$$

By means of the Maya diagrams, the Jacobi identity is rewritten as

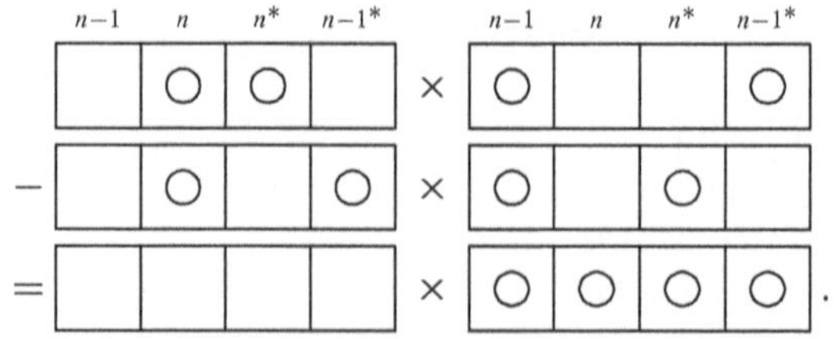

(2) As will be shown later, if one expresses the solutions to the KP equation and the Toda lattice equation as grammian determinants, their bilinear equations become Jacobi identities. As a matter of fact, the author did not know of the Jacobi identity when he started his research into solitons. He expected that the solution to the bilinear KdV equation could be expressed as a determinant but could not prove it in that form. As a result of trial and error, he expanded the determinant, changed the way of writing the solution, and finally managed the proof by establishing a particular polynomial identity (the Hirota condition). Luckily, this expression and the proof are effective for all types (A-type, B-type, ... [15]) of soliton equations. The proof by means of the Jacobi identity is useless for the verification of the solutions to B-type soliton equations, as these are expressed as pfaffians not determinants. Therefore, if the author had known the Jacobi identity, the discovery of B-type soliton equations might have taken much longer. □

2.7 Special determinants

2.7.1 Perfect square formula (i)

Employing the Jacobi identity, we may prove the theorem which is used in the definition of a pfaffian. Namely, that the determinant of the $n \times n$ antisymmetric matrix $A_n = \det[a_{j,k}]_{1 \le j,k \le n}$ is equal to zero if n is odd and equal to a perfect square of a polynomial in $a_{j,k}$, called the pfaffian, if n is even.

By using elementary row operations, we may show that $A_n = (-1)^n A_n$. If n is odd, then $A_n = -A_n$, and hence $A_n = 0$. For example,

$$\begin{vmatrix} 0 & a_{1,2} & a_{1,3} \\ -a_{1,2} & 0 & a_{2,3} \\ -a_{1,3} & -a_{2,3} & 0 \end{vmatrix} = \begin{vmatrix} 0 & -a_{1,2} & -a_{1,3} \\ a_{1,2} & 0 & -a_{2,3} \\ a_{1,3} & a_{2,3} & 0 \end{vmatrix} = -\begin{vmatrix} 0 & a_{1,2} & a_{1,3} \\ -a_{1,2} & 0 & a_{2,3} \\ -a_{1,3} & -a_{2,3} & 0 \end{vmatrix}, \tag{2.84}$$

and hence $A_3 = 0$.

Let D be the determinant of a $2m \times 2m$ antisymmetric matrix. Then we have

$$D\begin{bmatrix}2m\\2m\end{bmatrix} = D\begin{bmatrix}2m-1\\2m-1\end{bmatrix} = 0, \quad D\begin{bmatrix}2m-1\\2m\end{bmatrix} = -D\begin{bmatrix}2m\\2m-1\end{bmatrix}. \tag{2.85}$$

Hence, the Jacobi identity gives

$$D\begin{bmatrix}2m-1 & 2m\\2m-1 & 2m\end{bmatrix} D = \left(D\begin{bmatrix}2m-1\\2m\end{bmatrix}\right)^2 \quad (m = 1, 2, 3, \ldots). \tag{2.86}$$

In the case $m = 1$, we have $D = a_{1,2}^2$. Using this recurrence relation, we deduce that D is a perfect square for arbitrary m.

2.7.2 Perfect square formula (ii)

Consider the $n \times n$ identity matrix E and antisymmetric $n \times n$ matrices A and B. We will show that the determinant $\det(E + AB)$ is a perfect square,

$$\det(E + AB) = (1, 2, \ldots, n, 1^*, 2^*, \ldots, n^*)^2, \tag{2.87}$$

where $A = (a_{i,j})_{1\le i,j\le n}$, $B = (b_{i,j})_{1\le i,j\le n}$ and the pfaffian entries are given by

$$\begin{aligned}(i, j) &= a_{i,j} = -a_{j,i},\\ (i^*, j^*) &= b_{i,j} = -b_{j,i},\\ (i, j^*) &= (j, i^*) = \delta_{i,j}.\end{aligned}$$

For example, in the case of $n = 3$, this gives

$$\begin{aligned}&\begin{vmatrix}1+a_{1,2}b_{2,1}+a_{1,3}b_{3,1} & a_{1,3}b_{3,2} & a_{1,2}b_{2,3}\\ a_{2,3}b_{3,1} & 1+a_{2,1}b_{1,2}+a_{2,3}b_{3,2} & a_{2,1}b_{1,3}\\ a_{3,2}b_{2,1} & a_{3,1}b_{1,2} & 1+a_{3,1}b_{1,3}+a_{3,2}b_{2,3}\end{vmatrix}\\ &= (1, 2, 3, 1^*, 2^*, 3^*)^2\\ &= [(1, 2)(3, 1^*, 2^*, 3^*) - (1, 3)(2, 1^*, 2^*, 3^*)\\ &\quad + (1, 1^*)(2, 3, 2^*, 3^*)]^2\end{aligned}$$

$$
\begin{aligned}
&= [(1,2)(3,3^*)(1^*,2^*) + (1,3)(2,2^*)(1^*,3^*) \\
&\quad + (1,1^*)(2,3)(2^*,3^*) - (1,1^*)(2,2^*)(3,3^*)]^2 \\
&= [a_{1,2}b_{1,2} + a_{1,3}b_{1,3} + a_{2,3}b_{2,3} - 1]^2.
\end{aligned}
\tag{2.88}
$$

The identity (2.87) is proved as follows. First of all, we express the square of the pfaffian on the right-hand side as the determinant of an antisymmetric matrix:

$$
(1,2,\cdots,n,1^*,2^*,\cdots,n^*)^2
$$

$$
= \begin{vmatrix}
(1,1) & (1,2) & (1,3) & \cdots & (1,n) & (1,1^*) & (1,2^*) & \cdots & (1,n^*) \\
(2,1) & (2,2) & (2,3) & \cdots & (2,n) & (2,1^*) & (2,2^*) & \cdots & (2,n^*) \\
\vdots & \vdots & \vdots & & \vdots & \vdots & \vdots & & \vdots \\
(n,1) & (n,2) & (n,3) & \cdots & (n,n) & (n,1^*) & (n,2^*) & \cdots & (n,n^*) \\
(1^*,1) & (1^*,2) & (1^*,3) & \cdots & (1^*,n) & (1^*,1^*) & (1^*,2^*) & \cdots & (1^*,n^*) \\
(2^*,1) & (2^*,2) & (2^*,3) & \cdots & (2^*,n) & (2^*,1^*) & (2^*,2^*) & \cdots & (2^*,n^*) \\
\vdots & \vdots & \vdots & & \vdots & \vdots & \vdots & & \vdots \\
(n^*,1) & (n^*,2) & (n^*,3) & \cdots & (n^*,n) & (n^*,1^*) & (n^*,2^*) & \cdots & (n^*,n^*)
\end{vmatrix}
$$

$$
= \begin{vmatrix}
0 & a_{1,2} & a_{1,3} & \cdots & a_{1,n} & 1 & 0 & 0 & \cdots & 0 \\
a_{2,1} & 0 & a_{2,3} & \cdots & a_{2,n} & 0 & 1 & 0 & \cdots & 0 \\
a_{3,1} & a_{3,2} & 0 & \cdots & a_{3,n} & 0 & 0 & 1 & \cdots & 0 \\
\vdots & \vdots & \vdots & \ddots & \vdots & \vdots & \vdots & \vdots & \ddots & \vdots \\
a_{n,1} & a_{n,2} & a_{n,3} & \cdots & 0 & 0 & 0 & 0 & \cdots & 1 \\
-1 & 0 & 0 & \cdots & 0 & 0 & b_{1,2} & b_{1,3} & \cdots & b_{1,n} \\
0 & -1 & 0 & \cdots & 0 & b_{2,1} & 0 & b_{2,3} & \cdots & b_{2,n} \\
0 & 0 & -1 & \cdots & 0 & b_{3,1} & b_{3,2} & 0 & \cdots & b_{3,n} \\
\vdots & \vdots & \vdots & \ddots & \vdots & \vdots & \vdots & \vdots & \ddots & \vdots \\
0 & 0 & 0 & \cdots & -1 & b_{n,1} & b_{n,2} & b_{n,3} & \cdots & 0
\end{vmatrix}.
$$

Interchanging columns of this determinant so that an identity matrix is on its diagonal and multiplying each of the final n rows by -1, the right-hand side equals

$$\begin{vmatrix} 1 & 0 & 0 & \cdots & 0 & 0 & a_{1,2} & a_{1,3} & \cdots & a_{1,n} \\ 0 & 1 & 0 & \cdots & 0 & a_{2,1} & 0 & a_{2,3} & \cdots & a_{2,n} \\ 0 & 0 & 1 & \cdots & 0 & a_{3,1} & a_{3,2} & 0 & \cdots & a_{3,n} \\ \vdots & \vdots & \vdots & \ddots & \vdots & \vdots & \vdots & \vdots & \ddots & \vdots \\ 0 & 0 & 0 & \cdots & 1 & a_{n,1} & a_{n,2} & a_{n,3} & \cdots & 0 \\ 0 & -b_{1,2} & -b_{1,3} & \cdots & -b_{1,n} & 1 & 0 & 0 & \cdots & 0 \\ -b_{2,1} & 0 & -b_{2,3} & \cdots & -b_{2,n} & 0 & 1 & 0 & \cdots & 0 \\ -b_{3,1} & -b_{3,2} & 0 & \cdots & -b_{3,n} & 0 & 0 & 1 & \cdots & 0 \\ \vdots & \vdots & \vdots & \ddots & \vdots & \vdots & \vdots & \vdots & \ddots & \vdots \\ -b_{n,1} & -b_{n,2} & -b_{n,3} & \cdots & 0 & 0 & 0 & 0 & \cdots & 1 \end{vmatrix}$$

$$= \begin{vmatrix} E & A \\ -B & E \end{vmatrix} = |E + AB|.$$

2.7.3 Bordered determinants

Bordered determinants play an important role in proving that the bilinear KP and two-dimensional Toda lattice equations can be expressed as Jacobi identities.

Let A be an $n \times n$ matrix, let $|A|$ be its determinant and let $\Delta_{i,j}$ be its cofactor with respect to some matrix entry $a_{i,j}$. Then it is quite easy to prove that [34]

$$\begin{vmatrix} a_{1,1} & a_{1,2} & a_{1,3} & \cdots & a_{1,n} & x_1 \\ a_{2,1} & a_{2,2} & a_{2,3} & \cdots & a_{2,n} & x_2 \\ a_{3,1} & a_{3,2} & a_{3,3} & \cdots & a_{3,n} & x_3 \\ \vdots & \vdots & \vdots & & \vdots & \vdots \\ a_{n,1} & a_{n,2} & a_{n,3} & \cdots & a_{n,n} & x_n \\ y_1 & y_2 & y_3 & \cdots & y_n & z \end{vmatrix} = |A|z - \sum_{i,j=1}^{n} \Delta_{i,j} x_i y_j. \qquad (2.89)$$

Expanding the determinant on the left-hand side with respect to the $(n+1)$th column, we have

$$|A|z + \sum_{i=1}^{n}(-1)^{n+i+1}\begin{vmatrix} a_{11} & a_{12} & a_{13} & \cdots & a_{1n} \\ \vdots & \vdots & \vdots & & \vdots \\ a_{i-1,1} & a_{i-1,2} & a_{i-1,3} & \cdots & a_{i-1,n} \\ i & & & & \\ a_{i+1,1} & a_{i+1,2} & a_{i+1,3} & \cdots & a_{i+1,n} \\ \vdots & \vdots & \vdots & & \vdots \\ a_{n1} & a_{n2} & a_{n3} & \cdots & a_{nn} \\ y_1 & y_2 & y_3 & \cdots & y_n \end{vmatrix} x_i .$$

Expanding each of the above determinants with respect to its nth row, we obtain

$$\begin{vmatrix} a_{1,1} & a_{1,2} & a_{1,3} & \cdots & a_{1,n} \\ \vdots & \vdots & \vdots & & \vdots \\ a_{i-1,1} & a_{i-1,2} & a_{i-1,3} & \cdots & a_{i-1,n} \\ i & & & & \\ a_{i+1,1} & a_{i+1,2} & a_{i+1,3} & \cdots & a_{i+1,n} \\ \vdots & \vdots & \vdots & & \vdots \\ a_{n,1} & a_{n,2} & a_{n,3} & \cdots & a_{n,n} \\ y_1 & y_2 & y_3 & \cdots & y_n \end{vmatrix}$$

$$= \sum_{j=1}^{n}(-1)^{n+j}\begin{vmatrix} a_{1,1} & \cdots & a_{1,j-1} & j & a_{1,j+1} & \cdots & a_{1,n} \\ \vdots & & \vdots & & \vdots & \vdots & \\ a_{i-1,1} & \cdots & a_{i-1,j-1} & & a_{i-1,j+1} & \cdots & a_{i-1,n} \\ i & & & & & & \\ a_{i+1,1} & \cdots & a_{i+1,j-1} & & a_{i+1,j+1} & \cdots & a_{i+1,n} \\ \vdots & & \vdots & & \vdots & \vdots & \\ a_{n,1} & \cdots & a_{n,j-1} & & a_{n,j+1} & \cdots & a_{n,n} \end{vmatrix} y_j,$$

from which we can observe the coefficient of $x_i y_j$ to be $-\Delta_{i,j}$.

By moving the $(n+1)$th row and column to the first row and column, respectively, this bordered determinant can also be written as

$$\begin{vmatrix} z & y_1 & y_2 & y_3 & \cdots & y_n \\ x_1 & a_{1,1} & a_{1,2} & a_{1,3} & \cdots & a_{1,n} \\ x_2 & a_{2,1} & a_{2,2} & a_{2,3} & \cdots & a_{2,n} \\ x_3 & a_{3,1} & a_{3,2} & a_{3,3} & \cdots & a_{3,n} \\ \vdots & \vdots & \vdots & \vdots & & \vdots \\ x_n & a_{n,1} & a_{n,2} & a_{n,3} & \cdots & a_{n,n} \end{vmatrix} = |A|z - \sum_{i,j=1}^{n} \Delta_{i,j} x_i y_j. \qquad (2.90)$$

Remarks

(1) If $z = 1$, the left-hand side of (2.90) is

$$\begin{vmatrix} 1 & y_1 & y_2 & y_3 & \cdots & y_n \\ x_1 & a_{1,1} & a_{1,2} & a_{1,3} & \cdots & a_{1,n} \\ x_2 & a_{2,1} & a_{2,2} & a_{2,3} & \cdots & a_{2,n} \\ x_3 & a_{3,1} & a_{3,2} & a_{3,3} & \cdots & a_{3,n} \\ \vdots & \vdots & \vdots & \vdots & & \vdots \\ x_n & a_{n,1} & a_{n,2} & a_{n,3} & \cdots & a_{n,n} \end{vmatrix}$$

$$= \begin{vmatrix} 1 & y_1 & y_2 & \cdots & y_n \\ 0 & a_{1,1} - x_1 y_1 & a_{1,2} - x_1 y_2 & \cdots & a_{1,n} - x_1 y_n \\ 0 & a_{2,1} - x_2 y_1 & a_{2,2} - x_2 y_2 & \cdots & a_{2,n} - x_2 y_n \\ 0 & a_{3,1} - x_3 y_1 & a_{3,2} - x_3 y_2 & \cdots & a_{3,n} - x_3 y_n \\ \vdots & \vdots & \vdots & & \vdots \\ 0 & a_{n,1} - x_n y_1 & a_{n,2} - x_n y_2 & \cdots & a_{n,n} - x_n y_n \end{vmatrix},$$

after using Gaussian elimination. Hence, for $z = 1$, (2.90) gives

$$\begin{vmatrix} a_{1,1} - x_1 y_1 & a_{1,2} - x_1 y_2 & \cdots & a_{1,n} - x_1 y_n \\ a_{2,1} - x_2 y_1 & a_{2,2} - x_2 y_2 & \cdots & a_{2,n} - x_2 y_n \\ a_{3,1} - x_3 y_1 & a_{3,2} - x_3 y_2 & \cdots & a_{3,n} - x_3 y_n \\ \vdots & \vdots & & \vdots \\ a_{n,1} - x_n y_1 & a_{n,2} - x_n y_2 & \cdots & a_{n,n} - x_n y_n \end{vmatrix} = |A| - \sum_{i,j=1}^{n} \Delta_{i,j} x_i y_j. \qquad (2.91)$$

If we impose the additional constraints that (i) the matrix entries are anti-symmetric, $a_{i,j} = -a_{j,i}$, and (ii) $x_i = y_i$, then $\Delta_{j,i} = -\Delta_{i,j}$, and so the second term on the right-hand side of (2.91) is

$$\sum_{i,j=1}^{n} \Delta_{i,j} y_i y_j = -\sum_{i,j=1}^{n} \Delta_{j,i} y_j y_i = -\sum_{i,j=1}^{n} \Delta_{i,j} y_i y_j = 0.$$

Therefore, if n is even, we obtain the identity

$$\begin{vmatrix} a_{1,1}-y_1y_1 & a_{1,2}-y_1y_2 & \cdots & a_{1,n}-y_1y_n \\ a_{2,1}-y_2y_1 & a_{2,2}-y_2y_2 & \cdots & a_{2,n}-y_2y_n \\ a_{3,1}-y_3y_1 & a_{3,2}-y_3y_2 & \cdots & a_{3,n}-y_3y_n \\ \vdots & \vdots & & \vdots \\ a_{n,1}-y_ny_1 & a_{n,2}-y_ny_2 & \cdots & a_{n,n}-y_ny_n \end{vmatrix}$$

$$= |A| = (1, 2, \cdots, n)^2, \tag{2.92}$$

where the pfaffian entries are $(i, j) = a_{i,j}$.

(2) Equation (2.92) implies that adding $y_i y_j$ to each entry $a_{i,j}$ of an antisymmetric matrix does not change the value of the pfaffian of $(a_{i,j})$. This fact will be used in Chapter 3 to derive the relationship between the solutions of the KP equation given by τ_{KP} and those of the BKP equation given by τ_{BKP},

$$\tau_{\mathrm{KP}} = \tau_{\mathrm{BKP}}^2.$$

□

The following identity holds for bordered determinants:

$$\begin{vmatrix} 0 & y_1 & y_2 & y_3 & \cdots & y_n \\ z_1 & a_{1,1} & a_{1,2} & a_{1,3} & \cdots & a_{1,n} \\ z_2 & a_{2,1} & a_{2,2} & a_{2,3} & \cdots & a_{2,n} \\ z_3 & a_{3,1} & a_{3,2} & a_{3,3} & \cdots & a_{3,n} \\ \vdots & \vdots & \vdots & \vdots & & \vdots \\ z_n & a_{n,1} & a_{n,2} & a_{n,3} & \cdots & a_{n,n} \end{vmatrix} - \begin{vmatrix} 0 & 0 & y_1 & y_2 & y_3 & \cdots & y_n \\ 0 & 0 & c_1 & c_2 & c_3 & \cdots & c_n \\ x_1 & b_1 & a_{1,1} & a_{1,2} & a_{1,3} & \cdots & a_{1,n} \\ x_2 & b_2 & a_{2,1} & a_{2,2} & a_{2,3} & \cdots & a_{2,n} \\ x_3 & b_3 & a_{3,1} & a_{3,2} & a_{3,3} & \cdots & a_{3,n} \\ \vdots & \vdots & \vdots & \vdots & \vdots & & \vdots \\ x_n & b_n & a_{n,1} & a_{n,2} & a_{n,3} & \cdots & a_{n,n} \end{vmatrix}$$

$$= \begin{vmatrix} 0 & y_1 & y_2 & y_3 & \cdots & y_n \\ z_1 & b_1c_1 & b_1c_2 & b_1c_3 & \cdots & b_1c_n \\ x_2 & a_{2,1} & a_{2,2} & a_{2,3} & \cdots & a_{2,n} \\ x_3 & a_{3,1} & a_{3,2} & a_{3,3} & \cdots & a_{3,n} \\ \vdots & \vdots & \vdots & \vdots & & \vdots \\ x_n & a_{n,1} & a_{n,2} & a_{n,3} & \cdots & a_{n,n} \end{vmatrix} +$$

$$
+\begin{vmatrix}
0 & y_1 & y_2 & y_3 & \cdots & y_n \\
x_1 & a_{1,1} & a_{1,2} & a_{1,3} & \cdots & a_{1,n} \\
z_2 & b_2c_1 & b_2c_2 & b_2c_3 & \cdots & b_2c_n \\
x_3 & a_{3,1} & a_{3,2} & a_{3,3} & \cdots & a_{3,n} \\
\vdots & \vdots & \vdots & \vdots & & \vdots \\
x_n & a_{n,1} & a_{n,2} & a_{n,3} & \cdots & a_{n,n}
\end{vmatrix}
$$

$$
+\begin{vmatrix}
0 & y_1 & y_2 & y_3 & \cdots & y_n \\
x_1 & a_{1,1} & a_{1,2} & a_{1,3} & \cdots & a_{1,n} \\
x_2 & a_{2,1} & a_{2,2} & a_{2,3} & \cdots & a_{2,n} \\
z_3 & b_3c_1 & b_3c_2 & b_3c_3 & \cdots & b_3c_n \\
\vdots & \vdots & \vdots & \vdots & & \vdots \\
x_n & a_{n,1} & a_{n,2} & a_{n,3} & \cdots & a_{n,n}
\end{vmatrix} + \cdots
$$

$$
+\begin{vmatrix}
0 & y_1 & y_2 & y_3 & \cdots & y_n \\
x_1 & a_{1,1} & a_{1,2} & a_{1,3} & \cdots & a_{1,n} \\
x_2 & a_{2,1} & a_{2,2} & a_{2,3} & \cdots & a_{2,n} \\
x_3 & a_{3,1} & a_{3,2} & a_{3,3} & \cdots & a_{3,n} \\
\vdots & \vdots & \vdots & \vdots & & \vdots \\
z_n & b_nc_1 & b_nc_2 & b_nc_3 & \cdots & b_nc_n.
\end{vmatrix}. \tag{2.93}
$$

The above identity is linear in $c_1, c_2, \ldots, c_n$ and can be proved by comparing the coefficients of c_j on each side for $j = 1, 2, \ldots, n$. First, we set $c_1 = c_2 = \cdots = c_n = 0$. Then (2.93) holds because expansion of its left-hand side by the first column gives its right-hand side. Next, considering the coefficient in c_1 of this identity we have

$$
\begin{vmatrix}
0 & 0 & y_2 & \cdots & y_n \\
x_1 & b_1 & a_{1,2} & \cdots & a_{1,n} \\
x_2 & b_2 & a_{2,2} & \cdots & a_{2,n} \\
x_3 & b_3 & a_{3,2} & \cdots & a_{3,n} \\
\vdots & \vdots & \vdots & & \vdots \\
x_n & b_n & a_{n,2} & \cdots & a_{n,n}
\end{vmatrix}
= b_1 \begin{vmatrix}
0 & y_2 & \cdots & y_n \\
x_2 & a_{2,2} & \cdots & a_{2,n} \\
x_3 & a_{3,2} & \cdots & a_{3,n} \\
\vdots & \vdots & & \vdots \\
x_n & a_{n,2} & \cdots & a_{n,n}
\end{vmatrix} -
$$

$$
- b_2 \begin{vmatrix} 0 & y_2 & \cdots & y_n \\ x_1 & a_{1,2} & \cdots & a_{1,n} \\ x_3 & a_{3,2} & \cdots & a_{3,n} \\ \vdots & \vdots & & \vdots \\ x_n & a_{n,2} & \cdots & a_{n,n} \end{vmatrix} + b_3 \begin{vmatrix} 0 & y_2 & \cdots & y_n \\ x_1 & a_{1,2} & \cdots & a_{1,n} \\ x_2 & a_{2,2} & \cdots & a_{2,n} \\ x_4 & a_{4,2} & \cdots & a_{4,n} \\ \vdots & \vdots & & \vdots \\ x_n & a_{n,2} & \cdots & a_{n,n} \end{vmatrix} - \cdots
$$

$$
+ (-1)^{n+1} b_n \begin{vmatrix} 0 & y_2 & \cdots & y_n \\ x_1 & a_{1,2} & \cdots & a_{1,n} \\ x_2 & a_{2,2} & \cdots & a_{2,n} \\ \vdots & \vdots & & \vdots \\ x_{n-1} & a_{n-1,2} & \cdots & a_{n-1,n} \end{vmatrix} .
$$

This holds because the right-hand side is simply the expansion of the left-hand side by the second column. Similarly, the identity holds for the coefficient of any c_j. This completes the proof of (2.93).

2.8 Pfaffian identities

There are various kinds of pfaffian identities which will be proved in this section. We start with the expansion formula for the mth-order pfaffian

$$
(a_1, a_2, a_3, \ldots, a_{2m}) = \sum_{j=2}^{2m} (-1)^j (a_1, a_j)(a_2, a_3, \ldots, \widehat{a}_j, \ldots, a_{2m}). \tag{2.94}
$$

Appending $2n$ indices $1, 2, \ldots, 2n$ homogeneously to each pfaffian in the above formula, we obtain an extended expansion formula,

$$
\begin{aligned}
&(a_1, a_2, \ldots, a_{2m}, 1, 2, \ldots, 2n)(1, 2, \ldots, 2n) \\
&= \sum_{j=2}^{2m} (-1)^j (a_1, a_j, 1, 2, \ldots, 2n) \\
&\quad \times (a_2, a_3, \ldots, \widehat{a}_j, \ldots, a_{2m}, 1, 2, \ldots, 2n).
\end{aligned} \tag{2.95}
$$

Next, expanding the vanishing pfaffian $(a_1, a_2, a_3, \ldots, a_m, 2n, 1, 1)$ (m is odd), with respect to the final index 1, we obtain

$$0 = \sum_{j=1}^{m} (-1)^{j-1} (a_1, a_2, a_3, \ldots, \widehat{a}_j, \ldots, a_m, 2n, 1)(a_j, 1) \\ - (a_1, a_2, a_3, \ldots, a_m, 1)(2n, 1).$$

Therefore, we have

$$(a_1, a_2, a_3, \ldots, a_m, 1)(1, 2n) = \sum_{j=1}^{m} (-1)^{j-1} (a_j, 1) \\ \times (a_1, a_2, a_3, \ldots, \widehat{a}_j, \ldots, a_m, 1, 2n).$$

Appending $2n-2$ indices $2, 3, \ldots, 2n-1$ homogeneously to each pfaffian again, we obtain the identity

$$(a_1, a_2, \ldots, a_m, 1, 2, \ldots, 2n-1)(1, 2, \ldots, 2n) \\ = \sum_{j=1}^{m} (-1)^{j-1} (a_j, 1, 2, \ldots, 2n-1) \\ \times (a_1, a_2, a_3, \ldots, \widehat{a}_j, \ldots, a_m, 1, 2, \ldots, 2n). \tag{2.96}$$

These formulae will be proved below.

For example, in the case $m = 2$, (2.95) is written as

$$(a_1, a_2, a_3, a_4, 1, 2, \ldots, 2n)(1, 2, \ldots, 2n) \\ = (a_1, a_2, 1, 2, \ldots, 2n)(a_3, a_4, 1, 2, \ldots, 2n) \\ - (a_1, a_3, 1, 2, \ldots, 2n)(a_2, a_4, 1, 2, \ldots, 2n) \\ + (a_1, a_4, 1, 2, \ldots, 2n)(a_2, a_3, 1, 2, \ldots, 2n). \tag{2.95'}$$

In the case $m = 3$, (2.96) is written as

$$(a_1, a_2, a_3, 1, 2, \ldots, 2n-1)(1, 2, \ldots, 2n) \\ = (a_1, 1, 2, \ldots, 2n-1)(a_2, a_3, 1, 2, \ldots, 2n) \\ - (a_2, 1, 2, \ldots, 2n-1)(a_1, a_3, 1, 2, \ldots, 2n) \\ + (a_3, 1, 2, \ldots, 2n-1)(a_1, a_2, 1, 2, \ldots, 2n). \tag{2.96'}$$

Remarks

(1) Equations (2.95′) and (2.96′) are expressed in terms of Maya diagrams as

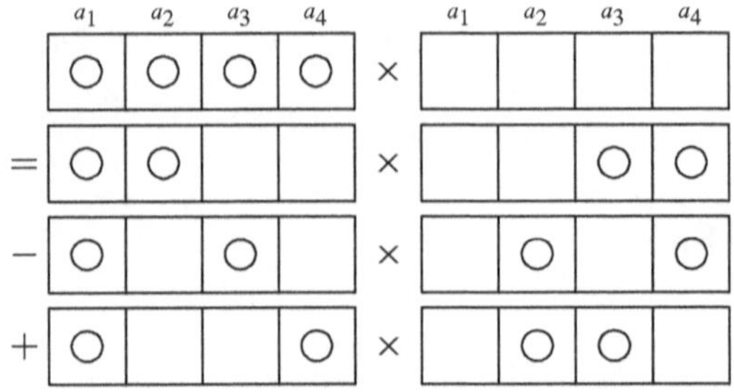

and

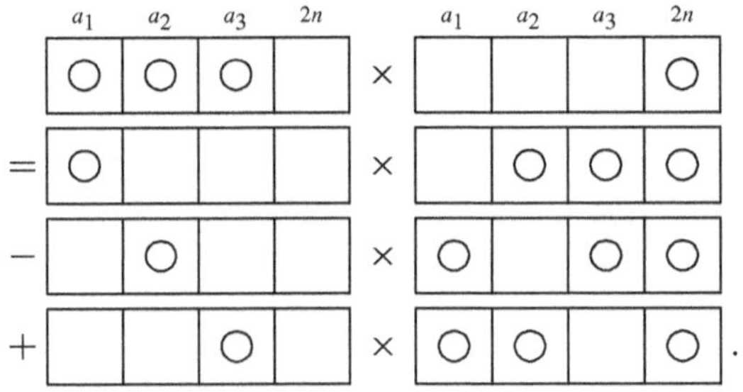

(2) One of main themes in the direct method is that a variety of soliton equations give rise to the above pfaffian identities. □

The rest of this section is devoted to proving these identities. We begin by proving the identity [35]

$$\sum_{j=0}^{M}(-1)^j(b_0, b_1, b_2, \ldots, \widehat{b_j}, \ldots, b_M)(b_j, c_0, c_1, \ldots, c_N)$$
$$= \sum_{k=0}^{N}(-1)^k(b_0, b_1, b_2, \ldots, b_M, c_k)(c_0, c_1, c_2, \ldots, \widehat{c_k}, \ldots, c_N). \tag{2.97}$$

This identity is expressed using Maya diagrams as

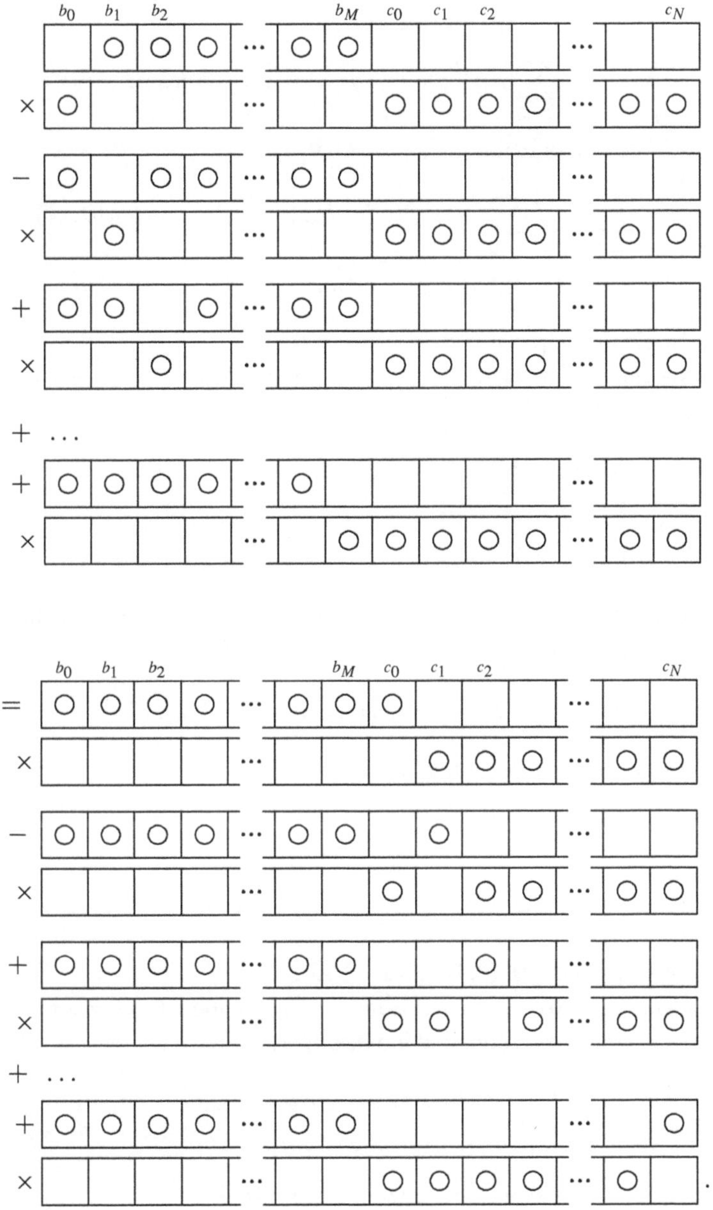

The proof of (2.97) is quite simple. Expanding pfaffians $(b_j, c_0, c_1, c_2, \ldots, c_N)$ on the left-hand side with respect to the first index b_j and $(b_0, b_1, b_2, \ldots, b_M, c_k)$ on the right-hand side with respect to the final index

c_k, gives

$$\sum_{j=0}^{M}(-1)^j \sum_{k=0}^{N}(-1)^k (b_0, b_1, b_2, \ldots, \widehat{b_j}, \ldots, b_M)(b_j, c_k)$$
$$\times (c_0, c_1, c_2, \ldots, \widehat{c_k}, \ldots, c_N)$$
$$= \sum_{k=0}^{N}(-1)^k \sum_{j=0}^{M}(-1)^j (b_0, b_1, b_2, \ldots, \widehat{b_j}, \ldots, b_M)(b_j, c_k)$$
$$\times (c_0, c_1, c_2, \ldots, \widehat{c_k}, \ldots, c_N),$$

which completes the proof.

As a special case of the identity (2.97), we select $M = 2n$, $N = 2m + 2n - 2$ and choose indices b_j, c_k to be

$$b_0 = a_1,\ b_1 = 1,\ b_2 = 2,\ \ldots,\ b_M = b_{2n} = 2n,$$
$$c_0 = a_2,\ c_1 = a_3,\ c_2 = a_4,\ \ldots,\ c_{2m-2} = a_{2m},$$
$$c_{2m-1} = 1,\ c_{2m} = 2,\ \ldots,\ c_N = c_{2m+2n-2} = 2n.$$

Since the above choice makes all except the $j = 0$ term on the left-hand side of (2.97) equal to zero, the left-hand side is

$$(1, 2, \ldots, 2n)(a_1, a_2, a_3, \ldots, a_{2m}, 1, 2, \ldots, 2n).$$

Also, the right-hand side is equal to

$$\sum_{k=0}^{2m-2} (-1)^k (a_1, 1, 2, \ldots, 2n, a_{k+2})$$
$$\times (a_2, a_3, \ldots, \widehat{a_{k+2}}, \ldots, a_{2m}, 1, 2, \ldots, 2n),$$

and so, for these choices, (2.97) gives (2.95).

The identity (2.96) is obtained from (2.95) by choosing $M = 2n - 2$, $N = m + 2n - 1$, where m is odd, and indices b_j, c_k to be

$$b_0 = 1,\ b_1 = 2,\ b_2 = 3,\ \ldots,\ b_M = b_{2n-2} = 2n - 1,$$
$$c_0 = a_1,\ c_1 = a_2,\ c_2 = a_3, \ldots,\ c_{m-1} = a_m,$$
$$c_m = 1,\ c_{m+1} = 2,\ \ldots,\ c_N = c_{m+2n-1} = 2n.$$

Remarks

(1) In the next chapter, we will show that bilinear soliton equations give rise to the fundamental pfaffian identities (2.95) and (2.96). It is mysterious

that the identities (2.95) and (2.96) are generated from the very simple identity (2.97).

(2) The Plücker relation has significance in projective geometry. However, the geometric meaning of the identity (2.97), which is obtained from a generalization of the Plücker relation, is still unknown. □

2.9 Expansion formulae for the pfaffian $(a_1, a_2, 1, 2, \ldots, 2n)$

If $(a_1, a_2) = 0$, the pfaffian $(a_1, a_2, 1, 2, \ldots, 2n)$ can be expanded in two different ways [31]:

$$\begin{aligned}(a_1, a_2, 1, 2, \ldots, 2n) = &\sum_{1 \le j < k \le 2n} (-1)^{j+k-1} (a_1, a_2, j, k) \\ &\times (1, 2, \ldots, \widehat{j}, \ldots, \widehat{k}, \ldots, 2n)\end{aligned} \tag{2.98}$$

or

$$\begin{aligned}(a_1, a_2, 1, 2, \ldots, 2n) = &\sum_{j=2}^{2n} (-1)^j \big[(a_1, a_2, 1, j)(2, 3, \ldots, \widehat{j}, \ldots, 2n) \\ &+ (1, j)(a_1, a_2, 2, 3, \ldots, \widehat{j}, \ldots, 2n)\big].\end{aligned} \tag{2.99}$$

Expansion formula (2.98) is proved simply by expanding the pfaffian $(a_1, a_2, 1, 2, \ldots, 2n)$ first with respect to a_1 and then a_2. We have

$$\begin{aligned}&(a_1, a_2, 1, 2, \ldots, 2n) \\ &\quad = \sum_{j=1}^{2n} \sum_{k=1}^{2n} (-1)^{j+k} (a_1, j)(a_2, k)(1, 2, \ldots, \widehat{j}, \ldots, \widehat{k}, \ldots, 2n) \\ &\quad = \sum_{1 \le j < k \le 2n} (-1)^{j+k} \left[(a_1, j)(a_2, k) - (a_1, k)(a_2, j)\right] \\ &\quad\quad \times (1, 2, \ldots, \widehat{j}, \ldots, \widehat{k}, \ldots, 2n) \\ &\quad = \sum_{1 \le j < k \le 2n} (-1)^{j+k-1} (a_1, a_2, j, k)(1, 2, \ldots, \widehat{j}, \ldots, \widehat{k}, \ldots, 2n),\end{aligned}$$

since $(a_1, a_2) = 0$.

In order to prove the expansion formula (2.99), one has only to expand the pfaffian $(a_1, a_2, 1, 2, \ldots, 2n)$ with respect to the index 1:

$$
\begin{aligned}
&(a_1, a_2, 1, 2, \ldots, 2n) \\
&\quad = (1, a_1)(a_2, 2, \ldots, 2n) - (1, a_2)(a_1, 2, \ldots, 2n) \\
&\qquad + \sum_{j=2}^{2n} (-1)^j (1, j)(a_1, a_2, 2, 3, \ldots, \widehat{j}, \ldots, 2n).
\end{aligned}
$$

Next, the pfaffians $(a_2, 2, \ldots, 2n)$ and $(a_1, 2, \ldots, 2n)$ are expanded to give

$$
\begin{aligned}
(a_1, a_2, 1, 2, \ldots, 2n) &= (1, a_1) \sum_{j=2}^{2n} (-1)^j (a_2, j)(2, 3, \ldots, \widehat{j}, \ldots, 2n) \\
&\quad - (1, a_2) \sum_{j=2}^{2n} (-1)^j (a_1, j)(2, 3, \ldots, \widehat{j}, \ldots, 2n) \\
&\quad + \sum_{j=2}^{2n} (-1)^j (1, j)(a_1, a_2, 2, 3, \ldots, \widehat{j}, \ldots, 2n).
\end{aligned}
$$

Making use of the condition $(a_1, a_2) = 0$, the right-hand side of the above is

$$
\begin{aligned}
&\sum_{j=2}^{2n} (-1)^j \left[(a_1, a_2, 1, j)(2, 3, \ldots, \widehat{j}, \ldots, 2n) \right. \\
&\qquad \left. + (1, j)(a_1, a_2, 2, 3, \ldots, \widehat{j}, \ldots, 2n) \right],
\end{aligned}
$$

as required.

Expansion formula (2.98) can be generalized by considering the pfaffian $(b_1, b_2, 1, 2, \ldots, 2n)$ instead of $(1, 2, \ldots, 2n)$. We have

$$
\begin{aligned}
(a_1, a_2, b_1, b_2, 1, 2, \ldots, 2n) &= \sum_{j=1}^{2n} \sum_{k=j+1}^{2n} (-1)^{j+k-1} (a_1, a_2, j, k) \\
&\times (b_1, b_2, 1, 2, 3, \ldots, \widehat{j}, \ldots, \widehat{k}, \ldots, 2n), \qquad (2.100)
\end{aligned}
$$

where $(a_j, b_k) = 0$ for $j, k = 1, 2$.

Remark

We will make use of these expansion formulae later in connection with the derivatives of pfaffians. □

2.10 Addition formulae for pfaffians

We consider an nth-order pfaffian $(1, 2, \ldots, 2n)_c$ with special entries $(i, j)_c$ given by sums of pfaffians,

$$(i, j)_c = (i, j) + \lambda(a, b, i, j). \tag{2.101}$$

In the above equation, λ is a parameter and we also suppose that

$$(a, b) = 0.$$

Then the following addition formula holds for an arbitrary n:

$$(1, 2, \ldots, 2n)_c = (1, 2, \ldots, 2n) + \lambda(a, b, 1, 2, \ldots, 2n). \tag{2.102}$$

For example, in the case $n = 2$, we have

$$\begin{aligned}(1, 2, 3, 4)_c &= (1, 2)_c(3, 4)_c - (1, 3)_c(2, 4)_c + (1, 4)_c(2, 3)_c \\ &= [(1, 2) + \lambda(a, b, 1, 2)]\,[(3, 4) + \lambda(a, b, 3, 4)] \\ &\quad - [(1, 3) + \lambda(a, b, 1, 3)]\,[(2, 4) + \lambda(a, b, 2, 4)] \\ &\quad + [(1, 4) + \lambda(a, b, 1, 4)]\,[(2, 3) + \lambda(a, b, 2, 3)] \\ &= (1, 2, 3, 4) + \lambda(a, b, 1, 2, 3, 4).\end{aligned} \tag{2.103}$$

Let us prove the addition formula (2.102) by induction. Obviously, the formula holds if $n = 1$. We suppose that the addition formula holds for an arbitrary $(n-1)$th-order pfaffian,

$$\begin{aligned}(2, 3, \ldots, \hat{j}, \ldots, 2n)_c &= (2, 3, \ldots, \hat{j}, \ldots, 2n) \\ &\quad + \lambda(a, b, 2, 3, \ldots, \hat{j}, \ldots, 2n).\end{aligned} \tag{2.104}$$

Expansion of the left-hand side of (2.102) gives

$$(1, 2, \ldots, 2n)_c = \sum_{j=2}^{2n} (-1)^j (1, j)_c (2, 3, \ldots, \hat{j}, \ldots, 2n)_c.$$

Employing (2.101) and (2.104), the right-hand side equals

$$\begin{aligned}&\sum_{j=2}^{2n} (-1)^j\, [(1, j) + \lambda(a, b, 1, j)] \\ &\qquad \times \left[(2, 3, \ldots, \hat{j}, \ldots, 2n) + \lambda(a, b, 2, 3, \ldots, \hat{j}, \ldots, 2n)\right].\end{aligned}$$

Here, the coefficient of λ^0 is obviously $(1, 2, \ldots, 2n)$ and the coefficient of λ^1 is $(a, b, 1, 2, \ldots, 2n)$ because of the expansion formula (2.99) in the previous section. We next investigate the coefficient of λ^2. In the expansion formula (2.99),

$$\begin{aligned}(a, b, 1, 2, \ldots, 2n, 2n+1, 2n+2) \\ = \sum_{j=2}^{2n+2} (-1)^j \big[(a, b, 1, j)(2, 3, \ldots, \widehat{j}, \ldots, 2n, 2n+1, 2n+2) \\ + (1, j)(a, b, 2, 3, \ldots, \widehat{j}, \ldots, 2n, 2n+1, 2n+2)\big],\end{aligned}$$

we set indices $2n + 1 = a$ and $2n + 2 = b$, so that we obtain

$$0 = \sum_{j=2}^{2n} (-1)^j (a, b, 1, j)(2, 3, \ldots, \widehat{j}, \ldots, 2n, a, b).$$

This shows that the coefficient of λ^2 is zero, and so we have

$$(1, 2, \ldots, 2n)_c = (1, 2, \ldots, 2n) + \lambda(a, b, 1, 2, \ldots, 2n),$$

and the proof of the addition formula (2.102) is complete.

By employing this pfaffian addition theorem, the determinantal identity (2.91),

$$\det |a_{i,j} - x_i y_j|_{1 \le i < j \le 2n} = \begin{vmatrix} a_{1,1} & a_{1,2} & \cdots & a_{1,n} \\ a_{2,1} & a_{2,2} & \cdots & a_{2,n} \\ \vdots & \vdots & & \vdots \\ a_{n,1} & a_{n,2} & \cdots & a_{n,n} \end{vmatrix} + \begin{vmatrix} 0 & y_1 & y_2 & \cdots & y_n \\ x_1 & a_{1,1} & a_{1,2} & \cdots & a_{1,n} \\ x_2 & a_{2,1} & a_{2,2} & \cdots & a_{2,n} \\ \vdots & \vdots & \vdots & & \vdots \\ x_n & a_{n,1} & a_{n,2} & \cdots & a_{n,n} \end{vmatrix}, \qquad (2.91')$$

may be rewritten as

$$\begin{aligned}(1, 2, \ldots, n, n^*, \ldots, 2^*, 1^*)_c = (1, 2, \ldots, n, n^*, \ldots, 2^*, 1^*) \\ + (d_0^*, d_0, 1, 2, \ldots, n, n^*, \ldots, 2^*, 1^*),\end{aligned} \qquad (2.105)$$

where the (nonzero) pfaffian entries are given by

$$(i, j^*)_c = (i, j^*) + (d_0^*, d_0, i, j^*) = a_{i,j} - x_i y_j,$$
$$(i, j^*) = a_{i,j}, \quad (d_0^*, i) = x_i, \quad (d_0, j^*) = y_j.$$

Remark

This addition theorem is useful for finding difference analogues of soliton equations and in finding difference formulae for pfaffians. □

2.11 Derivative formulae for pfaffians

The x-derivative of an arbitrary determinant $A \equiv \det(a_{i,j})_{1\le i,j\le n}$ is given by

$$\frac{\partial A}{\partial x} = \sum_{1\le i,j\le n} \frac{\partial a_{i,j}}{\partial x}\frac{\partial A}{\partial a_{i,j}} = \sum_{1\le i,j\le n} \frac{\partial a_{i,j}}{\partial x}\Delta_{i,j}. \tag{2.106}$$

We will derive here an analogous formula for the derivative of a pfaffian.

Consider the antisymmetric matrix whose entries (i, j) are the entries in the pfaffian $(1, 2, \ldots, 2n)$. Its determinant,

$$D = \det\big((i, j)\big)_{1\le i,j\le 2n}, \tag{2.107}$$

is given by $D = (1, 2, \ldots, 2n)^2$. From (2.106), the derivative of D with respect to x is

$$\frac{\partial D}{\partial x} = \sum_{1\le i,j\le 2n} \frac{\partial (i, j)}{\partial x}\Delta_{i,j}. \tag{2.108}$$

Substituting $D = (1, 2, \ldots, 2n)^2$ and using (2.68), that is

$$\Delta_{i,j} = \Gamma(i, j)\,(1, 2, \ldots, 2n),$$

we have

$$\frac{\partial}{\partial x}(1, 2, \ldots, 2n) = \frac{1}{2}\sum_{1\le i,j\le 2n} \frac{\partial (i, j)}{\partial x}\Gamma(i, j)$$
$$= \sum_{1\le i<j\le 2n} \frac{\partial (i, j)}{\partial x}\Gamma(i, j). \tag{2.109}$$

This is the formula for the derivative of a pfaffian.

If the x-derivative of a pfaffian entry (i, j) is expressed in terms of another pfaffian, for example

$$\frac{\partial (i, j)}{\partial x} = (a_0, b_0, i, j), \quad (a_0, b_0) = 0, \tag{2.110}$$

we obtain

$$\frac{\partial}{\partial x}(1, 2, \ldots, 2n) = (a_0, b_0, 1, 2, \ldots, 2n) \tag{2.111}$$

by using the expansion formula (2.98) [31].

The above derivative formula may also be obtained without using the formula for the derivative of an antisymmetric determinant. We first differentiate directly the pfaffian expansion formula,

$$(1, 2, \ldots, 2n) = \sum_{j=1}^{2n} (-1)^{j-1} (1, j)(2, 3, \ldots, \widehat{j}, \ldots, 2n), \tag{2.112}$$

to obtain

$$\begin{aligned}\frac{\partial}{\partial x}(1, 2, \ldots, 2n) = \sum_{j=1}^{2n} (-1)^{j-1} \Big[&\frac{\partial}{\partial x}(1, j)(2, 3, \ldots, \widehat{j}, \ldots, 2n) \\ &+ (1, j)\frac{\partial}{\partial x}(2, 3, \ldots, \widehat{j}, \ldots, 2n)\Big];\end{aligned}$$

then we use induction. If $n = 1$, then (2.111) is simply (2.110). Under the assumption that the expansion formula holds for pfaffians of order $n - 1$, the right-hand side equals

$$\begin{aligned}&\sum_{j=1}^{2n} (-1)^{j-1} \big[(a_0, b_0, 1, j)(2, 3, \ldots, \widehat{j}, \ldots, 2n) \\ &\quad + (1, j)(a_0, b_0, 2, 3, \ldots, \widehat{j}, \ldots, 2n)\big] = (a_0, b_0, 1, 2, \ldots, 2n)\end{aligned}$$

using expansion formula (2.99). This concludes the proof of (2.111).

Next, let us calculate the derivative with respect to another variable y, say, of the pfaffian $(a_0, b_0, 1, 2, \ldots, 2n)$. There are many possible forms for the y-derivative of a pfaffian entry. For simplicity, we consider

$$\begin{aligned}\frac{\partial}{\partial y}(i, j) &\equiv (a_1, b_1, i, j), \\ \frac{\partial}{\partial y}(a_0, b_0, i, j) &\equiv (a_2, b_0, i, j) + (a_0, b_2, i, j),\end{aligned} \tag{2.113}$$

where $(a_i, a_j) = (a_i, b_j) = (b_i, b_j) = 0$ for $i, j = 0, 1, 2$. We will show that

$$\begin{aligned}&\frac{\partial}{\partial y}(a_0, b_0, 1, 2, \ldots, 2n)\\&\quad = (a_2, b_0, 1, 2, \ldots, 2n) + (a_0, b_2, 1, 2, \ldots, 2n)\\&\qquad + (a_0, b_0, a_1, b_1, 1, 2, \ldots, 2n).\end{aligned} \tag{2.114}$$

In order to prove this, let us consider a y-derivative of the expansion formula (2.98), that is

$$\begin{aligned}(a_0, b_0, 1, 2, \ldots, 2n) = &\sum_{1\leq i<j\leq 2n} (-1)^{i+j-1}(a_0, b_0, i, j)\\&\times (1, 2, \ldots, \widehat{i}, \ldots, \widehat{j}, \ldots, 2n),\end{aligned}$$

with derivative

$$\begin{aligned}&\frac{\partial}{\partial y}(a_0, b_0, 1, 2, \ldots, 2n)\\&\quad = \sum_{1\leq i<j\leq 2n} (-1)^{i+j-1}\big\{ [(a_2, b_0, i, j)\\&\qquad + (a_0, b_2, i, j)]\,(1, 2, \ldots, \widehat{i}, \ldots, \widehat{j}, \ldots, 2n)\\&\qquad + (a_0, b_0, i, j)(a_1, b_1, 1, 2, \ldots, \widehat{i}, \ldots, \widehat{j}, \ldots, 2n)\big\}.\end{aligned}$$

Using expansion formulae (2.99) and (2.100), the right-hand side equals

$$\begin{aligned}&(a_2, b_0, 1, 2, \ldots, 2n) + (a_0, b_2, 1, 2, \ldots, 2n)\\&\quad + (a_0, b_0, a_1, b_1, 1, 2, \ldots, 2n),\end{aligned}$$

which completes the proof.

As an application of the above formulae, let us calculate the derivative of the Nth-order determinant [36]

$$\tau_N = \det(a_{i,j})_{1\leq i,j\leq N}, \tag{2.115}$$

where

$$a_{i,j} \equiv c_{i,j} + \int^x f_i g_j \,\mathrm{d}x, \quad c_{i,j} = \text{constant},$$

and functions f_i, g_j $(i, j = 1, 2, \ldots, N)$ satisfy the linear differential equations

$$\frac{\partial}{\partial x_n} f_i = \frac{\partial^n}{\partial x^n} f_i, \quad \frac{\partial}{\partial x_n} g_j = (-1)^{n-1}\frac{\partial^n}{\partial x^n} g_j. \tag{2.116}$$

Remark

The determinant τ_N is one expression for the N-soliton solution of the KP equation. □

Since the calculation of derivatives of the determinant τ_N plays an important role in soliton theory, we will describe the methods for calculating the derivatives using both determinantal and pfaffian formulae. From the determinantal formula

$$\frac{\partial \tau_N}{\partial x} = \sum_{1\le i,j\le N} \frac{\partial a_{i,j}}{\partial x} \Delta_{i,j}, \tag{2.117}$$

we have

$$\frac{\partial \tau_N}{\partial x} = \sum_{1\le i,j\le N} f_i g_j \Delta_{i,j}. \tag{2.118}$$

We observe that the right-hand side of (2.118) may be written as a bordered determinant by putting $z = 0$ in (2.90), that is

$$\begin{vmatrix} z & y_1 & y_2 & y_3 & \cdots & y_n \\ x_1 & a_{1,1} & a_{1,2} & a_{1,3} & \cdots & a_{1,n} \\ x_2 & a_{2,1} & a_{2,2} & a_{2,3} & \cdots & a_{2,n} \\ x_3 & a_{3,1} & a_{3,2} & a_{3,3} & \cdots & a_{3,n} \\ \vdots & \vdots & \vdots & \vdots & & \vdots \\ x_n & a_{n,1} & a_{n,2} & a_{n,3} & \cdots & a_{n,n} \end{vmatrix} = |A|z - \sum_{i,j=1}^{n} \Delta_{ij} x_i y_j,$$

to obtain

$$\frac{\partial \tau_N}{\partial x} = - \begin{vmatrix} 0 & g_1 & g_2 & g_3 & \cdots & g_N \\ f_1 & a_{1,1} & a_{1,2} & a_{1,3} & \cdots & a_{1,N} \\ f_2 & a_{2,1} & a_{2,2} & a_{2,3} & \cdots & a_{2,N} \\ f_3 & a_{3,1} & a_{3,2} & a_{3,3} & \cdots & a_{3,N} \\ \vdots & \vdots & \vdots & \vdots & & \vdots \\ f_N & a_{N,1} & a_{N,2} & a_{N,3} & \cdots & a_{N,N} \end{vmatrix}. \tag{2.119}$$

As remarked in Section 2.4, an nth-order determinant can be expressed as an nth-order pfaffian. Making use of this fact, let us rewrite τ_N in terms of the pfaffian with entries given by

$$(i, j^*) = a_{i,j} \equiv c_{i,j} + \int^x f_i g_j \, \mathrm{d}x, \quad (i, j) = (i^*, j^*) = 0. \tag{2.120}$$

Then the determinant τ_N may be written as

$$\tau_N = (1, 2, \ldots, N, N^*, \ldots, 2^*, 1^*). \tag{2.121}$$

The x-derivative of pfaffian entry (i, j^*) is given by

$$\frac{\partial(i, j^*)}{\partial x} = f_i g_j = (d_0, d_0^*, i, j^*), \tag{2.122}$$

where

$$\begin{gathered}(d_n, j^*) = \frac{\partial^n}{\partial x^n} g_j, \quad (d_m, d_n^*) = 0, \quad (d_n^*, i) = \frac{\partial^n}{\partial x^n} f_i, \\ (d_n, i) = (d_m^*, j^*) = 0, \quad m, n = 0, 1, 2, \ldots . \end{gathered} \tag{2.123}$$

This pfaffian has very similar properties to a wronskian.

From the pfaffian derivative formula (2.111), we have

$$\frac{\partial}{\partial x}\tau_N = (d_0, d_0^*, 1, 2, \ldots, N, N^*, \ldots, 2^*, 1^*). \tag{2.124}$$

Re-expressing this in terms of determinants we have

$$\begin{aligned}\frac{\partial}{\partial x}\tau_N &= \begin{vmatrix} (d_0, d_0^*) & (d_0, 1^*) & \cdots & (d_0, N^*) \\ (1, d_0^*) & (1, 1^*) & \cdots & (1, N^*) \\ \vdots & \vdots & & \vdots \\ (N, d_0^*) & (N, 1^*) & \cdots & (N, N^*) \end{vmatrix} \\ &= \begin{vmatrix} 0 & g_1 & g_2 & \cdots & g_N \\ -f_1 & a_{1,1} & a_{1,2} & \cdots & a_{1,N} \\ -f_2 & a_{2,1} & a_{2,2} & \cdots & a_{2,N} \\ \vdots & \vdots & \vdots & & \vdots \\ -f_N & a_{N,1} & a_{N,2} & \cdots & a_{N,N} \end{vmatrix},\end{aligned} \tag{2.125}$$

in agreement with (2.119).

Next we calculate the second-order derivative $\tau_{N,xx}$. Recalling that $\partial a_{i,j}/\partial x = f_i g_j$, and using elementary properties of determinants, we have the following:

$$
\begin{aligned}
&\tau_{N,xx} \\
&= -\frac{\partial}{\partial x}\begin{vmatrix} 0 & g_1 & g_2 & \cdots & g_N \\ f_1 & a_{1,1} & a_{1,2} & \cdots & a_{1,N} \\ f_2 & a_{2,1} & a_{2,2} & \cdots & a_{2,N} \\ \vdots & \vdots & \vdots & & \vdots \\ f_N & a_{N,1} & a_{N,2} & \cdots & a_{N,N} \end{vmatrix} \\
&= -\begin{vmatrix} 0 & g_{1,x} & g_{2,x} & \cdots & g_{N,x} \\ f_1 & a_{1,1} & a_{1,2} & \cdots & a_{1,N} \\ f_2 & a_{2,1} & a_{2,2} & \cdots & a_{2,N} \\ \vdots & \vdots & \vdots & & \vdots \\ f_N & a_{N,1} & a_{N,2} & \cdots & a_{N,N} \end{vmatrix} - \begin{vmatrix} 0 & g_1 & g_2 & \cdots & g_N \\ f_{1,x} & 0 & 0 & \cdots & 0 \\ f_2 & a_{2,1} & a_{2,2} & \cdots & a_{2,N} \\ \vdots & \vdots & \vdots & & \vdots \\ f_N & a_{N,1} & a_{N,2} & \cdots & a_{N,N} \end{vmatrix} \\
&\quad - \cdots - \begin{vmatrix} 0 & g_1 & g_2 & \cdots & g_N \\ f_1 & a_{1,1} & a_{1,2} & \cdots & a_{1,N} \\ f_2 & a_{2,1} & a_{2,2} & \cdots & a_{2,N} \\ \vdots & \vdots & \vdots & & \vdots \\ f_{N,x} & 0 & 0 & \cdots & 0 \end{vmatrix}.
\end{aligned}
$$

By employing the expansion formula (2.93) for a bordered determinant, the last N terms on the right-hand side may be combined into a single determinant. Finally, we obtain the derivative formula

$$
\begin{aligned}
\tau_{N,xx} = &-\begin{vmatrix} 0 & g_{1,x} & g_{2,x} & \cdots & g_{N,x} \\ f_1 & a_{1,1} & a_{1,2} & \cdots & a_{1,N} \\ f_2 & a_{2,1} & a_{2,2} & \cdots & a_{2,N} \\ \vdots & \vdots & \vdots & & \vdots \\ f_N & a_{N,1} & a_{N,2} & \cdots & a_{N,N} \end{vmatrix} \\
&-\begin{vmatrix} 0 & g_1 & g_2 & \cdots & g_N \\ f_{1,x} & a_{1,1} & a_{1,2} & \cdots & a_{1,N} \\ f_{2,x} & a_{2,1} & a_{2,2} & \cdots & a_{2,N} \\ \vdots & \vdots & \vdots & & \vdots \\ f_{N,x} & a_{N,1} & a_{N,2} & \cdots & a_{N,N} \end{vmatrix}.
\end{aligned} \tag{2.126}
$$

On the other hand, in the case of the pfaffian expression, the derivatives of the entries of $\tau_{N,x}$ are

$$\begin{aligned}\frac{\partial}{\partial x}(d_0, d_0^*, i, j^*) = \frac{\partial}{\partial x} f_i g_j &= f_{ix} g_j + f_i g_{jx} \\ &= (d_1, d_0^*, i, j^*) + (d_0, d_1^*, i, j^*).\end{aligned}$$

Using the expansion formula (2.99), we obtain

$$\begin{aligned}\tau_{N,xx} &= \frac{\partial}{\partial x}(d_0, d_0^*, 1, 2, \ldots, N, N^*, \ldots, 2^*, 1^*) \\ &= (d_1, d_0^*, 1, 2, \ldots, N, N^*, \ldots, 2^*, 1^*) \\ &\quad + (d_0, d_1^*, 1, 2, \ldots, N, N^*, \ldots, 2^*, 1^*).\end{aligned}$$

Expressing the above result in terms of determinants gives

$$\begin{aligned}\tau_{N,xx} &= \begin{vmatrix} 0 & g_{1,x} & g_{2,x} & \cdots & g_{N,x} \\ -f_1 & a_{1,1} & a_{1,2} & \cdots & a_{1,N} \\ -f_2 & a_{2,1} & a_{2,2} & \cdots & a_{2,N} \\ \vdots & \vdots & \vdots & & \vdots \\ -f_N & a_{N,1} & a_{N,2} & \cdots & a_{N,N} \end{vmatrix} \\ &\quad + \begin{vmatrix} 0 & g_1 & g_2 & \cdots & g_N \\ -f_{1,x} & a_{1,1} & a_{1,2} & \cdots & a_{1,N} \\ -f_{2,x} & a_{2,1} & a_{2,2} & \cdots & a_{2,N} \\ \vdots & \vdots & \vdots & & \vdots \\ -f_{N,x} & a_{N,1} & a_{N,2} & \cdots & a_{N,N} \end{vmatrix}, \end{aligned} \tag{2.126$'$}$$

which coincides with the result obtained in the case of a determinant, (2.126).

Through a similar calculation, we obtain

$$\begin{aligned}\frac{\partial^3 \tau_N}{\partial x^3} &= -\begin{vmatrix} 0 & g_{1,xx} & \cdots & g_{N,xx} \\ f_1 & a_{1,1} & \cdots & a_{1,N} \\ \vdots & \vdots & & \vdots \\ f_N & a_{N,1} & \cdots & a_{N,N} \end{vmatrix} - 2\begin{vmatrix} 0 & g_{1,x} & \cdots & g_{N,x} \\ f_{1,x} & a_{1,1} & \cdots & a_{1,N} \\ \vdots & \vdots & & \vdots \\ f_{N,x} & a_{N,1} & \cdots & a_{N,N} \end{vmatrix} \\ &\quad - \begin{vmatrix} 0 & g_1 & \cdots & g_N \\ f_{1,xx} & a_{1,1} & \cdots & a_{1,N} \\ \vdots & \vdots & & \vdots \\ f_{N,xx} & a_{N,1} & \cdots & a_{N,N} \end{vmatrix},\end{aligned}$$

but to calculate $\tau_{N,xxxx}$ needs another result. Using the identity (2.93) for bordered determinants, we have

$$\begin{vmatrix} 0 & g_{1,x} & \cdots & g_{N,x} \\ f_{1,xx} & f_1 g_1 & \cdots & f_1 g_N \\ \vdots & \vdots & & \vdots \\ f_{N,x} & a_{N,1} & \cdots & a_{N,N} \end{vmatrix} + \begin{vmatrix} 0 & g_{1,x} & \cdots & g_{N,x} \\ f_{1,x} & a_{1,1} & \cdots & a_{1,N} \\ \vdots & \vdots & & \vdots \\ f_{N,x} & a_{N,1} & \cdots & a_{N,N} \end{vmatrix}$$
$$+ \cdots + \begin{vmatrix} 0 & g_{1,x} & \cdots & g_{N,x} \\ f_{1,x} & a_{1,1} & \cdots & a_{1,N} \\ \vdots & \vdots & & \vdots \\ f_{N,xx} & f_N g_1 & \cdots & f_N g_N \end{vmatrix}$$
$$= \begin{vmatrix} 0 & g_{1,x} & \cdots & g_{Nx} \\ f_{1,xx} & a_{1,1} & \cdots & a_{1,N} \\ \vdots & \vdots & & \vdots \\ f_{N,xx} & a_{N,1} & \cdots & a_{N,N} \end{vmatrix} - \begin{vmatrix} 0 & 0 & g_{1,x} & \cdots & g_{N,x} \\ 0 & 0 & g_1 & \cdots & g_N \\ f_{1,x} & f_1 & a_{1,1} & \cdots & a_{1,N} \\ \vdots & \vdots & \vdots & & \vdots \\ f_{N,x} & f_N & a_{N,1} & \cdots & a_{N,N} \end{vmatrix}.$$

This allows us to calculate that

$$\frac{\partial^4 \tau_N}{\partial x^4}$$
$$= - \begin{vmatrix} 0 & g_{1,xxx} & \cdots & g_{N,xxx} \\ f_1 & a_{1,1} & \cdots & a_{1,N} \\ \vdots & \vdots & & \vdots \\ f_N & a_{N,1} & \cdots & a_{N,N} \end{vmatrix} - 3 \begin{vmatrix} 0 & g_{1,xx} & \cdots & g_{N,xx} \\ f_{1,x} & a_{1,1} & \cdots & a_{1,N} \\ \vdots & \vdots & & \vdots \\ f_{N,x} & a_{N,1} & \cdots & a_{N,N} \end{vmatrix}$$
$$+ 2 \begin{vmatrix} 0 & 0 & g_{1,x} & \cdots & g_{N,x} \\ 0 & 0 & g_1 & \cdots & g_N \\ f_{1,x} & f_1 & a_{1,1} & \cdots & a_{1,N} \\ \vdots & \vdots & \vdots & & \vdots \\ f_{N,x} & f_N & a_{N,1} & \cdots & a_{N,N} \end{vmatrix} - 3 \begin{vmatrix} 0 & g_{1,x} & \cdots & g_{N,x} \\ f_{1,xx} & a_{1,1} & \cdots & a_{1,N} \\ \vdots & \vdots & & \vdots \\ f_{N,xx} & a_{N,1} & \cdots & a_{N,N} \end{vmatrix}$$
$$- \begin{vmatrix} 0 & g_1 & \cdots & g_N \\ f_{1,xxx} & a_{1,1} & \cdots & a_{1,N} \\ \vdots & \vdots & & \vdots \\ f_{N,xxx} & a_{N,1} & \cdots & a_{N,N} \end{vmatrix}. \tag{2.127}$$

On the other hand, in the case of the pfaffian expression for τ_N, we have

$$\begin{aligned}
\frac{\partial^4 \tau_N}{\partial x^4} &= \frac{\partial}{\partial x}[(d_2, d_0^*, 1, 2, \dots, N, N^*, \dots, 2^*, 1^*) \\
&\quad + 2(d_1, d_1^*, 1, 2, \dots, N, N^*, \dots, 2^*, 1^*) \\
&\quad + (d_0, d_2^*, 1, 2, \dots, N, N^*, \dots, 2^*, 1^*)] \\
&= (d_3, d_0^*, 1, 2, \dots, N, N^*, \dots, 2^*, 1^*) \\
&\quad + 3(d_2, d_1^*, 1, 2, \dots, N, N^*, \dots, 2^*, 1^*) \\
&\quad + 2(d_0, d_0^*, d_1, d_1^*, 1, 2, \dots, N, N^*, \dots, 2^*, 1^*) \\
&\quad + 3(d_1, d_2^*, 1, 2, \dots, N, N^*, \dots, 2^*, 1^*) \\
&\quad + (d_0, d_3^*, 1, 2, \dots, N, N^*, \dots, 2^*, 1^*),
\end{aligned} \tag{2.127'}$$

which again coincides with the result obtained in the case of a determinant.

Remark

As we have shown, the calculation of derivatives using pfaffians is much simpler than using determinants. We have presented the calculations using determinants for the benefit of those who are not familiar with pfaffians. □

3

Structure of soliton equations

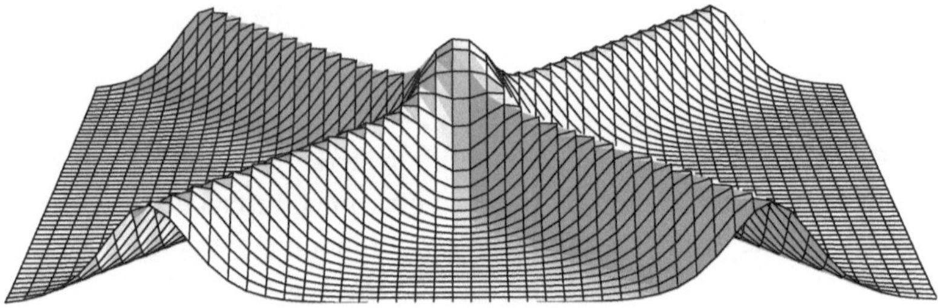

Head-on interaction.

3.1 Introduction

Mikio Sato [12, 13] was the first to discover that the KP (Kadomtsev–Petviashvili) equation is the most fundamental among the many soliton equations. Sato discovered that polynomial solutions of the bilinear KP equation are equivalent to the characteristic polynomials of the general linear group. Later, he found a Lax pair for a hierarchy of KP-like equations by means of a pseudo-differential operator, and came to the conclusion that the KP equation is equivalent to the motion of a point in a Grassmanian manifold and its bilinear equation is nothing but a Plücker relation. Also, Junkichi Satsuma [37] had discovered before Sato that the soliton solutions of the KdV equation could be expressed in terms of wronskian determinants. Later, in 1983, Freeman and Nimmo [38, 39] found that the KP bilinear equation could be rewritten as a determinantal identity if one expresses its soliton solutions in terms of wronskians. In this chapter, we develop the above results and show that some bilinear soliton equations having solutions expressed as pfaffians (or as determinants) are nothing but pfaffian identities.

Remark

The KdV equation is a $1+1$-dimensional equation describing shallow water waves. The KP equation was introduced in order to discuss the stability of these waves to perpendicular horizontal perturbations [40]. Some physicists may object to the idea that the KP equation is the most fundamental of the soliton equations. First of all, the KdV equation is derived through a certain approximation, and then the KP equation is obtained from the KdV equation under the assumption that horizontal perturbations are small. This means that the KP equation is far from a basic equation in the physical sense. However, the central issue here is not so much the physical viewpoint as the mathematical one. The KP equation is fundamental because of the simple mathematical structure of its solutions and its relation to the other soliton equations arising from this simplicity. □

3.2 The KP equation

3.2.1 Wronskian solutions

The KP equation is the 2+1-dimensional (two-dimensional space, (x, y) plus one-dimensional time, t) nonlinear partial differential equation

$$(u_t + u_{xxx} + 6uu_x)_x + u_{yy} = 0. \tag{3.1}$$

Since we obtain the KdV equation by neglecting the y-derivative term, this is also called the two-dimensional KdV equation.

We first present a wronskian expression for its solutions and observe how the KP equation reduces to a Plücker relation. The KP equation is equivalent to

$$(-4u_t + u_{xxx} + 6uu_x)_x + 3u_{yy} = 0 \tag{3.2}$$

under an appropriate scale transformation. Later, we will see the reason for this scaling of independent variables t, x and y. The dependent variable transformation

$$u = 2(\log \tau)_{xx} \tag{3.3}$$

gives the bilinear equation

$$(D_1^4 - 4D_1D_3 + 3D_2^2)\tau \cdot \tau = 0, \tag{3.4}$$

where we have rewritten the independent variables as

$$x = x_1, \quad y = x_2, \quad t = x_3,$$

and their derivatives as

$$D_x = D_1, \quad D_y = D_2, \quad D_t = D_3.$$

We have also chosen the constant of integration to be zero.

Following the procedure described in Chapter 1, the perturbation method gives the two-soliton solution

$$\tau_2 = 1 + \exp(\eta_1) + \exp(\eta_2) + a_{12}\exp(\eta_1 + \eta_2), \tag{3.5}$$

where $\eta_i = P_i x_1 + Q_i x_2 + \Omega_i x_3 + \eta_i^0$. The dispersion relation is given by

$$P_i^4 - 4P_i\Omega_i + 3Q_i^2 = 0 \quad (i = 1, 2), \tag{3.6}$$

and the phase shift term is given by

$$a_{12} = -\frac{(P_1 - P_2)^4 - 4(P_1 - P_2)(\Omega_1 - \Omega_2) + 3(Q_1 - Q_2)^2}{(P_1 + P_2)^4 - 4(P_1 + P_2)(\Omega_1 + \Omega_2) + 3(Q_1 + Q_2)^2}. \tag{3.7}$$

By putting $a_{ij} = \exp(A_{ij})$, the N-soliton solution is expressed as

$$\tau_N = \sum \exp\left[\sum_{i=1}^{N} \mu_i \eta_i + \sum_{i<j}^{(N)} A_{ij}\mu_i\mu_j\right], \tag{3.8}$$

where $\sum$ is a summation over all possible combinations of $\mu_1 = 0, 1,\ \mu_2 = 0, 1, \ldots,\ \mu_N = 0, 1$, and $\sum_{i<j}^{(N)}$ is the sum over all pairs i, j, where $i < j$, chosen from $\{1, 2, \ldots, N\}$.

Remark

The proof of the N-soliton solution τ_N of the KP equation, found by Satsuma [41], is very complicated compared with the case of the KdV equation. It is strange that, if we consider things from the perspective that bilinear equations are nothing but determinantal identities (Plücker relations) or, equivalently, if τ_N is expressed as a wronskian, the proof of the N-soliton solution is much easier for the KP equation than for the KdV equation. Later, we will describe wronskian and grammian expressions for the N-soliton solution. □

Let us introduce new parameters p_i, q_i for $i = 1, 2, \ldots$ [42], where

$$\begin{aligned} P_i &= p_i - q_i, \\ Q_i &= p_i^2 - q_i^2, \\ \Omega_i &= p_i^3 - q_i^3. \end{aligned} \tag{3.9}$$

Using this parametrization, the dispersion relation is satisfied automatically since

$$P_i^4 - 4P_i\Omega_i + 3Q_i^2 = (p_i - q_i)^4 - 4(p_i - q_i)(p_i^3 - q_i^3) + 3(p_i^2 - q_i^2)^2 = 0. \tag{3.10}$$

The phase shift term is rewritten more simply as

$$a_{12} = -\frac{(P_1 - P_2)^4 - 4(P_1 - P_2)(\Omega_1 - \Omega_2) + 3(Q_1 - Q_2)^2}{(P_1 + P_2)^4 - 4(P_1 + P_2)(\Omega_1 + \Omega_2) + 3(Q_1 + Q_2)^2} = \frac{(p_1 - p_2)(q_1 - q_2)}{(p_1 - q_2)(q_1 - p_2)}. \tag{3.11}$$

The scaling of t, x and y was chosen so as to obtain (3.10) and (3.11). We next rewrite η_i as

$$\eta_i = P_i x_1 + Q_i x_2 + \Omega_i x_3 + \text{constant} = \xi_i - \widehat{\xi}_i, \tag{3.12}$$

where

$$\xi_i = p_i x_1 + p_i^2 x_2 + p_i^3 x_3 + \text{constant},$$
$$\widehat{\xi}_i = q_i x_1 + q_i^2 x_2 + q_i^3 x_3 + \text{constant}.$$

Employing this notation, we have

$$\tau_2 = 1 + \exp(\xi_1 - \widehat{\xi}_1) + \exp(\xi_2 - \widehat{\xi}_2) + \frac{(p_1 - p_2)(q_1 - q_2)}{(p_1 - q_2)(q_1 - p_2)} \times \exp(\xi_1 + \xi_2 - \widehat{\xi}_1 - \widehat{\xi}_2). \tag{3.13}$$

We now define functions f_1 and f_2 by

$$f_1 = \exp(\xi_1) + \exp(\widehat{\xi}_1), \quad f_2 = \exp(\xi_2) + \exp(\widehat{\xi}_2), \tag{3.14}$$

and consider their wronskian,

$$\mathrm{Wr}(f_1, f_2) \equiv \begin{vmatrix} f_1 & f_{1,x} \\ f_2 & f_{2,x} \end{vmatrix} = \left[\exp(\xi_1) + \exp(\widehat{\xi}_1)\right]\left[p_2 \exp(\xi_2) + q_2 \exp(\widehat{\xi}_2)\right] - \left[p_1 \exp(\xi_1) + q_1 \exp(\widehat{\xi}_1)\right]\left[\exp(\xi_2) + \exp(\widehat{\xi}_2)\right] =$$

$$= (p_2 - p_1)\exp(\xi_1 + \xi_2) + (q_2 - p_1)\exp(\xi_1 + \widehat{\xi}_2) + (p_2 - q_1)\exp(\widehat{\xi}_1 + \xi_2) + (q_2 - q_1)\exp(\widehat{\xi}_1 + \widehat{\xi}_2), \tag{3.15}$$

where we have rewritten $x_1 = x$ for simplicity.

Dividing (3.15) by $(q_2 - q_1)\exp(\widehat{\xi}_1 + \widehat{\xi}_2)$, we have

$$\mathrm{Wr}(f_1, f_2) = (q_2 - q_1)\exp(\widehat{\xi}_1 + \widehat{\xi}_2)\left[1 + \frac{(q_2 - p_1)}{(q_2 - q_1)}\exp(\xi_1 - \widehat{\xi}_1) + \frac{(p_2 - q_1)}{(q_2 - q_1)}\exp(\xi_2 - \widehat{\xi}_2) + \frac{(p_2 - p_1)}{(q_2 - q_1)}\exp(\xi_1 + \xi_2 - \widehat{\xi}_1 - \widehat{\xi}_2)\right].$$

Then, using the fact that the solution u given by (3.3) is invariant under the transformation

$$\tau \to c_0 \exp(c_1 x_1 + c_2 x_2 + c_3 x_3 + \cdots)\tau, \tag{3.16}$$

where $c_0, c_1, c_2, \ldots$ are constants, we see that the wronskian solution is equivalent to

$$1 + \frac{(q_2 - p_1)}{(q_2 - q_1)}\exp(\xi_1 - \widehat{\xi}_1) + \frac{(p_2 - q_1)}{(q_2 - q_1)}\exp(\xi_2 - \widehat{\xi}_2) + \frac{(p_2 - p_1)}{(q_2 - q_1)}\exp(\xi_1 + \xi_2 - \widehat{\xi}_1 - \widehat{\xi}_2).$$

Next, we introduce phase constants $\delta_i, \widehat{\delta}_i$ defined by the relations

$$\begin{aligned} \frac{(q_2 - p_1)}{(q_2 - q_1)}\exp(\xi_1 - \widehat{\xi}_1) &= \exp(\xi_1 + \delta_1 - \widehat{\xi}_1 - \widehat{\delta}_1), \\ \frac{(p_2 - q_1)}{(q_2 - q_1)}\exp(\xi_2 - \widehat{\xi}_2) &= \exp(\xi_2 + \delta_2 - \widehat{\xi}_2 - \widehat{\delta}_2), \end{aligned} \tag{3.17}$$

and further replace variables $\xi_i, \widehat{\xi}_i$ as follows:

$$\xi_i + \delta_i \to \xi_i, \quad \widehat{\xi}_i + \widehat{\delta}_i \to \widehat{\xi}_i. \tag{3.18}$$

As a result, we see that $\mathrm{Wr}(f_1, f_2)$ can be rewritten as

$$\mathrm{Wr}(f_1, f_2) \propto 1 + \exp(\xi_1 - \widehat{\xi}_1) + \exp(\xi_2 - \widehat{\xi}_2) + \frac{(p_1 - p_2)(q_1 - q_2)}{(p_1 - q_2)(q_1 - p_2)} \times \exp(\xi_1 + \xi_2 - \widehat{\xi}_1 - \widehat{\xi}_2), \tag{3.19}$$

which is nothing but the two-soliton solution τ_2 found by the perturbation method.

Generalizing the above result, we might expect to be able to express the N-soliton solution as the $N \times N$ wronskian [12, 13, 38, 39]

$$\tau_N = \begin{vmatrix} f_1^{(0)} & f_1^{(1)} & \cdots & f_1^{(N-1)} \\ f_2^{(0)} & f_2^{(1)} & \cdots & f_2^{(N-1)} \\ \vdots & \vdots & & \vdots \\ f_N^{(0)} & f_N^{(1)} & \cdots & f_N^{(N-1)} \end{vmatrix}, \tag{3.20}$$

where $f_i^{(m)}$ is defined by

$$f_i^{(m)} = \frac{\partial^m f_i}{\partial x^m} \tag{3.21}$$

and each f_i $(i = 1, 2, \dots)$ satisfies the differential equations

$$\frac{\partial f_i}{\partial x_m} = \frac{\partial^m f_i}{\partial x^m}. \tag{3.22}$$

In order to confirm this, we only have to show that τ_N satisfies the bilinear equation [38, 39]

$$\begin{aligned}
&(D_1^4 - 4D_1D_3 + 3D_2^2)\tau \cdot \tau \\
&\quad = 2[\tau_{xxxx}\tau - 4\tau_{xxx}\tau_x + 3\tau_{xx}^2 - 4(\tau_{x_3x}\tau - \tau_{x_3}\tau_x) \\
&\qquad + 3(\tau_{x_2x_2}\tau - \tau_{x_2}^2)] \\
&\quad = 2[(\tau_{xxxx} - 4\tau_{x_3x} + 3\tau_{x_2x_2})\tau - 4(\tau_{xxx} - \tau_{x_3})\tau_x \\
&\qquad + 3(\tau_{xx} - \tau_{x_2})(\tau_{xx} + \tau_{x_2})] \\
&\quad = 0.
\end{aligned} \tag{3.23}$$

The derivative with respect to x of the wronskian

$$\tau_N = \begin{vmatrix} f_1^{(0)} & f_1^{(1)} & \cdots & f_1^{(N-1)} \\ f_2^{(0)} & f_2^{(1)} & \cdots & f_2^{(N-1)} \\ \vdots & \vdots & & \vdots \\ f_N^{(0)} & f_N^{(1)} & \cdots & f_N^{(N-1)} \end{vmatrix}$$

is equal to the sum of determinants, for $i = 1, 2, \dots, N$, in which the ith column of τ_N is replaced by its derivative. However, the derivative of the first column is equal to the second, the derivative of the second one equals the third, and so on. As a consequence, only the determinant with the last column differentiated remains. That is, the derivative $\tau_{N,x}$ is given by

$$\tau_{N,x} = \begin{vmatrix} f_1^{(0)} & f_1^{(1)} & \cdots & f_1^{(N-2)} & f_1^{(N)} \\ f_2^{(0)} & f_2^{(1)} & \cdots & f_2^{(N-2)} & f_2^{(N)} \\ \vdots & \vdots & & & \vdots \\ f_N^{(0)} & f_N^{(1)} & \cdots & f_N^{(N-2)} & f_N^{(N)} \end{vmatrix}. \tag{3.24}$$

This is one merit of the wronskian expression for τ_N; if we differentiate τ_N, only the number of derivatives in each column can change, the rows are unaffected. Therefore, we may adopt the simple notation

$$\begin{aligned} \tau_N &= [0, 1, \ldots, N-1] = \tau, \\ \tau_{N,x} &= [0, 1, \ldots, N-2, N]. \end{aligned} \tag{3.25}$$

These are expressed more simply by using the following Maya diagrams:

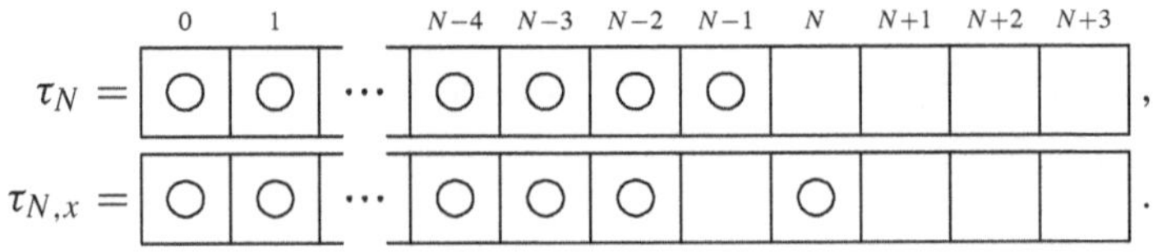

As discussed in Chapter 2, Maya diagrams were first introduced by Mikio Sato. In the language of physics, the diagram for τ_N represents the vacuum state in which fermions occupy cells $0, 1, 2, \ldots, N-2, N-1$. In the same way, the diagram for $\tau_{N,x}$ represents a state in which a fermion occupying the $(N-1)$th cell is excited into the Nth cell. From now on, for simplicity of notation, we write just τ for τ_N. Using the relation $f_{i,x_2} = f_{i,xx}$, the derivative of τ_N with respect to x_2 is

$$\begin{aligned} \tau_{x_2} &= [0, 1, \ldots, N-3, N-2, N+1] + [0, 1, \ldots, N-3, N, N-1] \\ &= [0, 1, \ldots, N-3, N-2, N+1] - [0, 1, \ldots, N-3, N-1, N]. \end{aligned} \tag{3.26}$$

The corresponding Maya diagram expression is

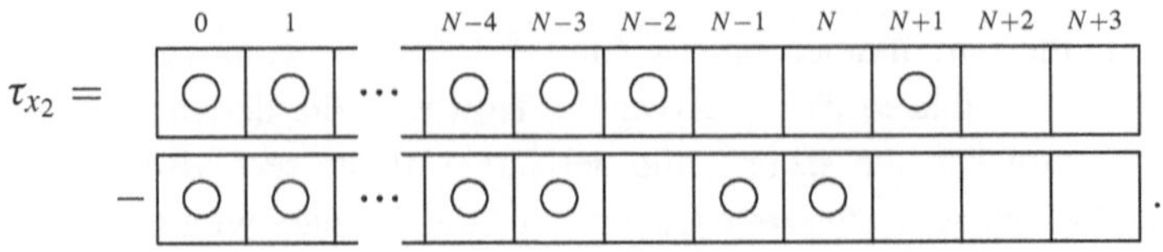

Remark

Each cell is occupied by at most one fermion. This corresponds to the fact that the determinant of a matrix in which two columns are equal is zero. □

Higher-order derivatives of τ_N are expressed, as excited fermion states, by means of Maya diagrams as follows [43]:

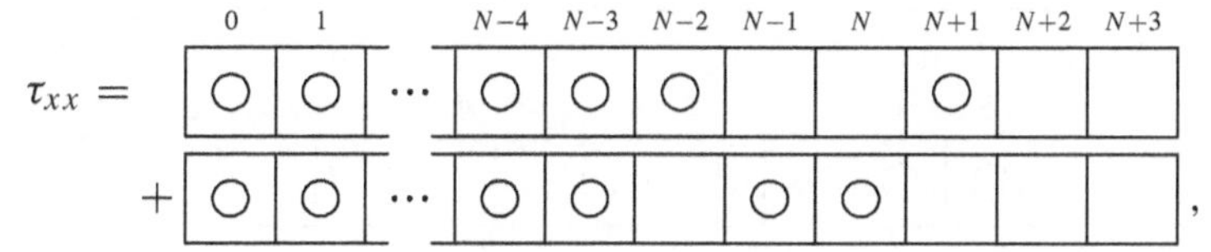

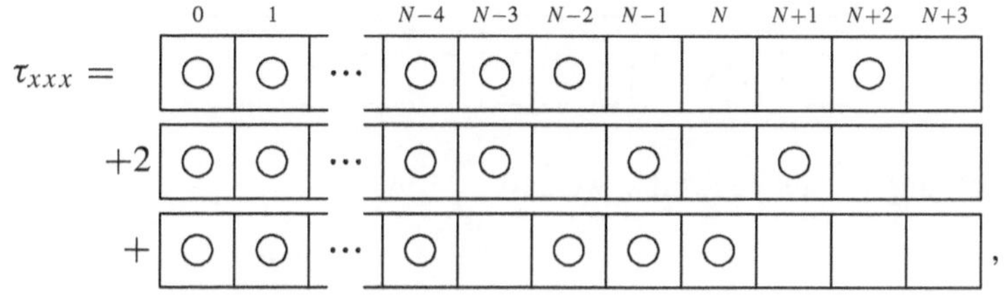

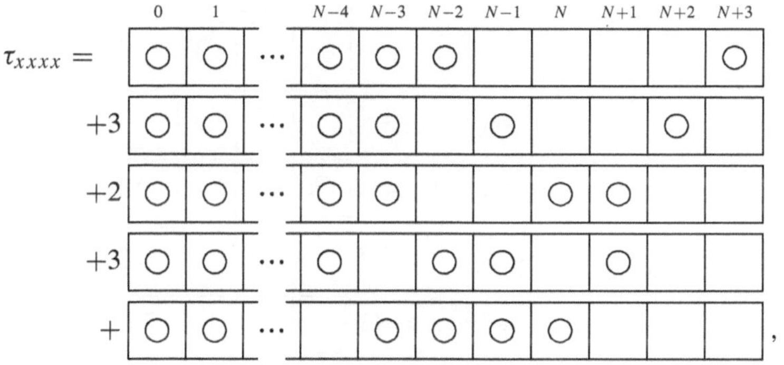

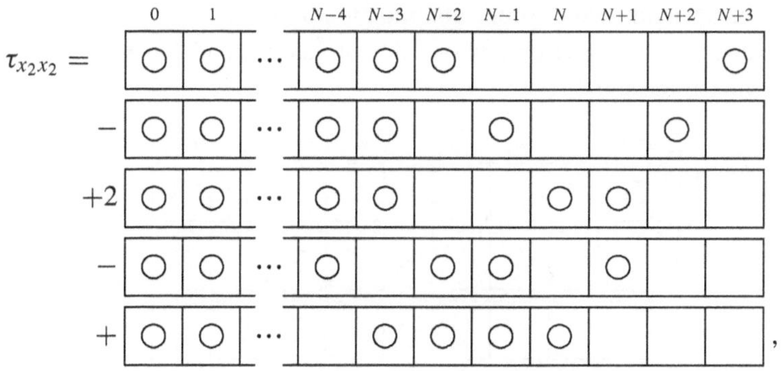

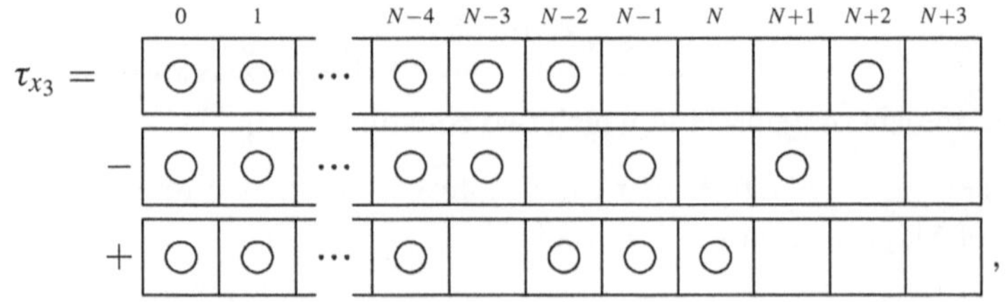

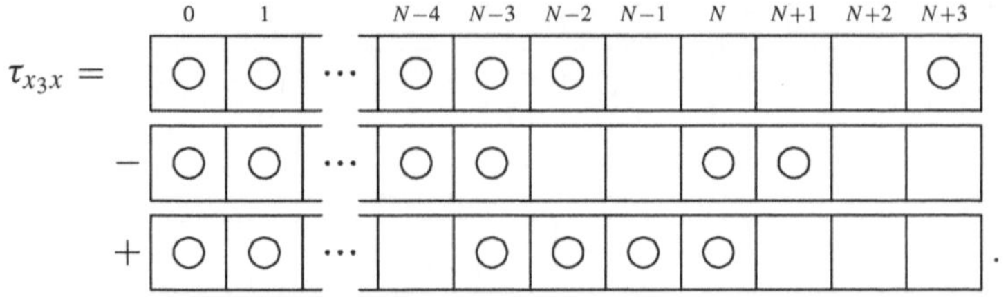

Summarizing the above results, we have

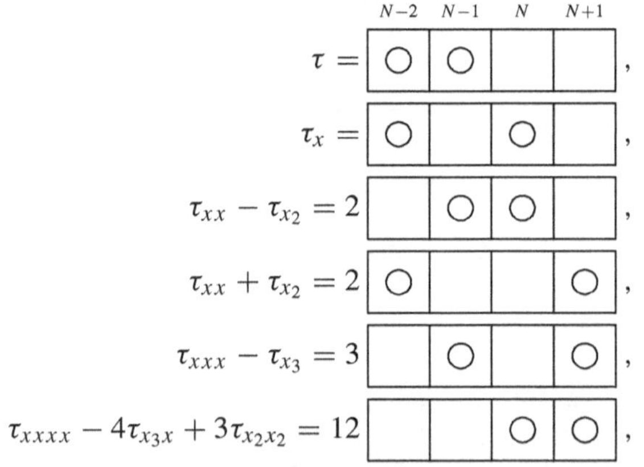

where we have omitted the cells common to all the Maya diagrams.

Substituting the above results into the bilinear form of the KP equation, we obtain the expression

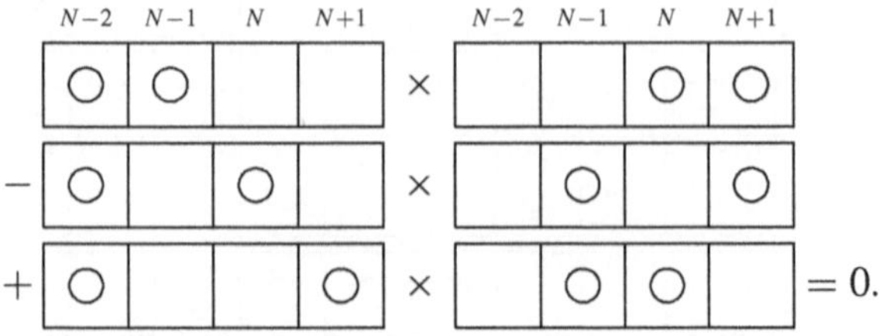

This is nothing but the Plücker relation for determinants [12, 13] (Section 2.5). Therefore, we have shown that $\tau = \tau_N$ satisfies the bilinear KP equation.

Remarks

(1) We have used a wronskian expression for τ. As we have shown in (2.29), a wronskian is expressed as a pfaffian,

$$\tau = (d_0, d_1, d_2, \ldots, d_{N-1}, 1, 2, \ldots, N), \tag{3.27}$$

where

$$(d_n, j) = \frac{\partial^n}{\partial x^n} f_j, \; (d_m, d_n) = 0 \; (j = 1, 2, \ldots, N, \; m, n = 0, 1, \ldots).$$

By employing this pfaffian expression, the pfaffian identity for the bilinear KP equation is expressed in terms of Maya diagrams as follows:

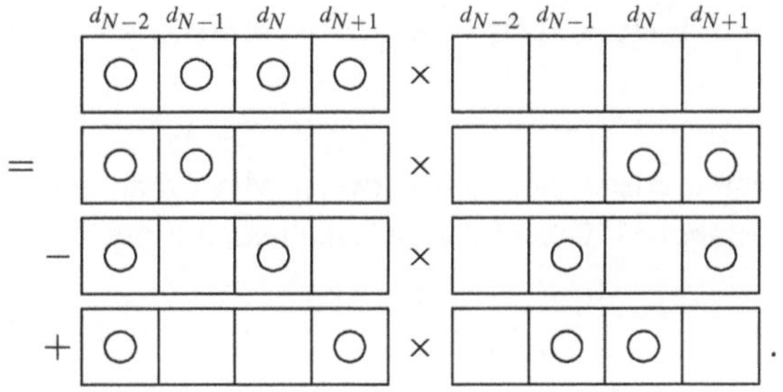

However, on the left-hand side of the above identity, the term

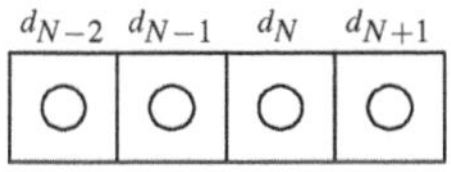

$= (d_0, d_1, d_2, \ldots, d_{N-1}, d_N, d_{N+1}, 1, 2, \ldots, N)$

is zero since the number of the symbols d_m is larger than number of symbols j, and

$$(d_m, d_n) = 0. \tag{3.28}$$

Hence, the pfaffian identity reduces to the Plücker relation.

(2) We have shown that τ solves the bilinear KP equation only by using the fact that the functions f_i, for $i = 1, 2, \ldots$, satisfy the linear differential equations

$$\frac{\partial f_i}{\partial x_m} = \frac{\partial^m f_i}{\partial x^m} \; (m = 1, 2, 3, \ldots). \tag{3.29}$$

Therefore, f_i can be chosen to be arbitrary solutions of (3.29). We note that such a function is obtained by taking any number of derivatives, with respect to p, of the exponential function

$$f = \exp(\xi), \ \xi = px_1 + p^2x_2 + p^3x_3 + p^4x_4 + \cdots .$$

(Characteristic polynomials, which are famous in group theory, are also contained in the above class of solutions [12, 13].)

(3) The function τ found here satisfies not only the KP equation, but also the following series of higher-order KP equations (the KP hierarchy) [15]:

$$\begin{aligned}
&(D_1^4 - 4D_1D_3 + 3D_2^2)\tau \cdot \tau = 0,\\
&[(D_1^3 + 2D_3)D_2 - 3D_1D_4]\tau \cdot \tau = 0,\\
&(D_1^6 - 20D_1^3D_3 - 80D_3^2 + 144D_1D_5 - 45D_1^2D_2^2)\tau \cdot \tau = 0,\\
&(D_1^6 + 4D_1^3D_3 - 32D_3^2 - 9D_1^2D_2^2 + 36D_2D_4)\tau \cdot \tau = 0,\\
&\cdots
\end{aligned} \tag{3.30}$$

(4) By using the correspondence between Maya diagrams and Young diagrams, the above Plücker relation is expressed using Young diagrams as

$$\tau_{\phi}\tau_{(2,2)} - \tau_{(1)}\tau_{(2,1)} + \tau_{(2)}\tau_{(1,1)} = 0.$$

□

Let us summarize the results obtained in this section. The KP equation,

$$(-4u_t + u_{xxx} + 6uu_x)_x + 3u_{yy} = 0, \tag{3.31}$$

is bilinearized to give

$$(D_1^4 - 4D_1D_3 + 3D_2^2)\tau \cdot \tau = 0, \tag{3.32}$$

through the dependent variable transformation

$$u = 2(\log \tau)_{xx}. \tag{3.33}$$

If we express the solution τ as a wronskian, the bilinear equation reduces to a Plücker relation and is represented as above in terms of Young diagrams. This fact was first discovered by Sato [12, 13].

3.2.2 Grammian solutions

We showed in Section 3.2.1 that the bilinear KP equation,

$$(D_1^4 - 4D_1D_3 + 3D_2^2)\tau \cdot \tau = 0, \tag{3.34}$$

is nothing but a Plücker relation if we express its solution τ as a wronskian. In this section, we show that this bilinear equation reduces to another type of determinantal identity, a Jacobi identity, by considering a different expression for τ.

As was shown in Section 3.2.1, the two-soliton solution τ_2 of the KP equation is written as

$$\begin{aligned}\tau_2 &= 1 + \exp(\xi_1 - \widehat{\xi}_1) + \exp(\xi_2 - \widehat{\xi}_2)\\ &\quad + \frac{(p_1 - p_2)(q_1 - q_2)}{(p_1 - q_2)(q_1 - p_2)} \exp(\xi_1 + \xi_2 - \widehat{\xi}_1 - \widehat{\xi}_2),\end{aligned} \tag{3.35}$$

where $\xi_i = p_i x_1 + p_i^2 x_2 + p_i^3 x_3 + \text{constant}$, and $\widehat{\xi}_i = q_i x_1 + q_i^2 x_2 + q_i^3 x_3 + \text{constant}$. By introducing phase factors δ_i and $\widehat{\delta}_j$ defined by

$$\exp(\xi_i - \widehat{\xi}_j) = \frac{1}{p_i - q_j} \exp(\xi_i + \delta_i - \widehat{\xi}_j - \widehat{\delta}_j), \tag{3.36}$$

and making the replacements

$$\xi_i + \delta_i \to \xi_i, \quad \widehat{\xi}_j + \widehat{\delta}_j \to \widehat{\xi}_j, \tag{3.37}$$

τ_2 may be rewritten as

$$\begin{aligned}\tau_2 &= 1 + \frac{1}{p_1 - q_1} \exp(\xi_1 - \widehat{\xi}_1) + \frac{1}{p_2 - q_2} \exp(\xi_2 - \widehat{\xi}_2)\\ &\quad + \alpha_{12} \exp(\xi_1 - \widehat{\xi}_1 + \xi_2 - \widehat{\xi}_2),\end{aligned} \tag{3.38}$$

where

$$\alpha_{12} = \frac{(p_1 - p_2)(q_1 - q_2)}{(p_1 - q_2)(q_1 - p_2)(p_1 - q_1)(p_2 - q_2)}.$$

This expression for τ_2 may be written as the determinant of a 2×2 matrix,

$$\tau_2 = \begin{vmatrix} 1 + \dfrac{1}{p_1 - q_1} \exp(\xi_1 - \widehat{\xi}_1) & \dfrac{1}{p_1 - q_2} \exp(\xi_1 - \widehat{\xi}_2) \\ \dfrac{1}{p_2 - q_1} \exp(\xi_2 - \widehat{\xi}_1) & 1 + \dfrac{1}{p_2 - q_2} \exp(\xi_2 - \widehat{\xi}_2) \end{vmatrix}; \tag{3.39}$$

that is, as

$$\tau_2 = \det(a_{ij})_{1\le i,j\le 2}, \quad a_{ij} = \delta_{ij} + \int^x \exp(\xi_i - \widehat{\xi}_j)\, \mathrm{d}x, \tag{3.40}$$

where $x = x_1$.

Remark

We choose the lower limit in the above integral to be $x = \pm\infty$ (the value of x such that the function $\exp(\xi_i - \widehat{\xi}_j)$ is equal to zero), but this is not an essential restriction. As can be seen later, the τ-function with the Kronecker delta δ_{ij} replaced by any constant matrix c_{ij} also solves the bilinear KP equation. Hence, we may choose the lower limit of the integral arbitrarily. □

Noting that the functions $\exp(\xi_i)$ and $\exp(-\widehat{\xi}_j)$ satisfy the linear differential equations

$$\begin{aligned} \frac{\partial}{\partial x_n} \exp \xi_i &= \frac{\partial^n}{\partial x^n} \exp \xi_i, \\ \frac{\partial}{\partial x_n} \exp(-\widehat{\xi}_j) &= (-1)^{n-1} \frac{\partial^n}{\partial x^n} \exp(-\widehat{\xi}_j), \end{aligned} \tag{3.41}$$

respectively, one might expect that the $N \times N$ determinant expression

$$\tau_N = \det(a_{ij})_{1\le i,j\le N}, \tag{3.42a}$$

$$a_{ij} = c_{ij} + \int^x f_i g_j \, \mathrm{d}x, \quad c_{ij} = \text{constant}, \tag{3.42b}$$

with f_i and g_j satisfying

$$\frac{\partial}{\partial x_n} f_i = \frac{\partial^n}{\partial x^n} f_i \quad (i = 1, 2, \dots, N), \tag{3.42c}$$

$$\frac{\partial}{\partial x_n} g_j = (-1)^{n-1} \frac{\partial^n}{\partial x^n} g_j \quad (j = 1, 2, \dots, N), \tag{3.42d}$$

is the N-soliton solution τ_N for the KP bilinear equation [44].

Remark

A grammian $G \equiv \det(g_{ij})_{1\le i,j\le N}$ is the determinant of a matrix with entries

$$g_{ij} \equiv \int_a^b f_i f_j \, \mathrm{d}x. \tag{3.43}$$

Because of the similarity between the above and the expression for τ_N, hereafter we also refer to the expression for τ_N as a grammian. □

In order to prove that this τ_N satisfies the bilinear equation

$$
\begin{aligned}
&(D_1^4 - 4D_1D_3 + 3D_2^2)\tau\cdot\tau \\
&\quad = 2[\tau_{xxxx}\tau - 4\tau_{xxx}\tau_x + 3\tau_{xx}^2 - 4(\tau_{x_3x}\tau - \tau_{x_3}\tau_x) \\
&\qquad + 3(\tau_{x_2x_2}\tau - \tau_{x_2}^2)] \\
&\quad = 2[(\tau_{xxxx} - 4\tau_{x_3x} + 3\tau_{x_2x_2})\tau - 4(\tau_{xxx} - \tau_{x_3})\tau_x \\
&\qquad + 3(\tau_{xx} - \tau_{x_2})(\tau_{xx} + \tau_{x_2})] \\
&\quad = 0,
\end{aligned}
\tag{3.44}
$$

let us determine the derivatives of the determinant τ_N. It is expressed by means of a pfaffian as

$$
\tau_N = (1, 2, \dots, N, N^*, \dots, 2^*, 1^*), \tag{3.45}
$$

where $(i, j^*) = c_{ij} + \int^x f_i g_j \, \mathrm{d}x$, c_{ij} = constant and $(i, j) = (i^*, j^*) = 0$.

Next let us introduce pfaffian entries

$$
\begin{aligned}
(d_n, j^*) &= \frac{\partial^n}{\partial x^n} g_j, \quad (d_m, d_n^*) = 0, \\
(d_n^*, i) &= \frac{\partial^n}{\partial x^n} f_i, \quad (d_n, i) = (d_m^*, i^*) = 0,
\end{aligned}
\tag{3.46}
$$

for $m, n = 0, 1, 2, 3, \dots$. In terms of these, derivatives of the elements $a_{ij} = (i, j^*)$ are given by

$$
\begin{aligned}
\frac{\partial}{\partial x} a_{ij} &= f_i g_j = (d_0, d_0^*, i, j^*), \\
\frac{\partial}{\partial x_2} a_{ij} &= \int^x (f_{ixx} g_j - f_i g_{jxx}) \, \mathrm{d}x \\
&= f_{ix} g_j - f_i g_{jx} \\
&= -(d_1, d_0^*, i, j^*) + (d_0, d_1^*, i, j^*), \\
\frac{\partial}{\partial x_3} a_{ij} &= \int^x (f_{ixxx} g_j + f_i g_{jxxx}) \, \mathrm{d}x \\
&= f_{ixx} g_j - f_{ix} g_{jx} + f_i g_{jxx} \\
&= (d_2, d_0^*, i, j^*) - (d_1, d_1^*, i, j^*) + (d_0, d_2^*, i, j^*).
\end{aligned}
\tag{3.47}
$$

Therefore, from the results in Section 2.11, we have

$$\begin{aligned}
\frac{\partial \tau_N}{\partial x} &= (d_0, d_0^*, 1, 2, \ldots, N, N^*, \ldots, 2^*, 1^*), \\
\frac{\partial^2 \tau_N}{\partial x^2} &= \frac{\partial}{\partial x}(d_0, d_0^*, 1, 2, \ldots, N, N^*, \ldots, 2^*, 1^*) \\
&= (d_1, d_0^*, 1, 2, \ldots, N, N^*, \ldots, 2^*, 1^*) \\
&\quad + (d_0, d_1^*, 1, 2, \ldots, N, N^*, \ldots, 2^*, 1^*), \\
\frac{\partial \tau_N}{\partial x_2} &= \frac{\partial}{\partial x_2}(1, 2, \ldots, N, N^*, \ldots, 2^*, 1^*) \\
&= -(d_1, d_0^*, 1, 2, \ldots, N, N^*, \ldots, 2^*, 1^*) \\
&\quad + (d_0, d_1^*, 1, 2, \ldots, N, N^*, \ldots, 2^*, 1^*), \\
\frac{\partial \tau_N}{\partial x_3} &= \frac{\partial}{\partial x_3}(1, 2, \ldots, N, N^*, \ldots, 2^*, 1^*) \\
&= (d_2, d_0^*, 1, 2, \ldots, N, N^*, \ldots, 2^*, 1^*) \\
&\quad - (d_1, d_1^*, 1, 2, \ldots, N, N^*, \ldots, 2^*, 1^*) \\
&\quad + (d_0, d_2^*, 1, 2, \ldots, N, N^*, \ldots, 2^*, 1^*), \\
\frac{\partial^3 \tau_N}{\partial x^3} &= \frac{\partial}{\partial x}\left[(d_1, d_0^*, 1, 2, \ldots, N, N^*, \ldots, 2^*, 1^*)\right. \\
&\quad \left. + (d_0, d_1^*, 1, 2, \ldots, N, N^*, \ldots, 2^*, 1^*)\right] \\
&= (d_2, d_0^*, 1, 2, \ldots, N, N^*, \ldots, 2^*, 1^*) \\
&\quad + 2(d_1, d_1^*, 1, 2, \ldots, N, N^*, \ldots, 2^*, 1^*) \\
&\quad + (d_0, d_2^*, 1, 2, \ldots, N, N^*, \ldots, 2^*, 1^*), \\
\frac{\partial^4 \tau_N}{\partial x^4} &= \frac{\partial}{\partial x}\left[(d_2, d_0^*, 1, 2, \ldots, N, N^*, \ldots, 2^*, 1^*)\right. \\
&\quad + 2(d_1, d_1^*, 1, 2, \ldots, N, N^*, \ldots, 2^*, 1^*) \\
&\quad \left. + (d_0, d_2^*, 1, 2, \ldots, N, N^*, \ldots, 2^*, 1^*)\right] \\
&= (d_3, d_0^*, 1, 2, \ldots, N, N^*, \ldots, 2^*, 1^*) \\
&\quad + 3(d_2, d_1^*, 1, 2, \ldots, N, N^*, \ldots, 2^*, 1^*) \\
&\quad + 2(d_0, d_0^*, d_1, d_1^*, 1, 2, \ldots, N, N^*, \ldots, 2^*, 1^*) \\
&\quad + 3(d_1, d_2^*, 1, 2, \ldots, N, N^*, \ldots, 2^*, 1^*) \\
&\quad + (d_0, d_3^*, 1, 2, \ldots, N, N^*, \ldots, 2^*, 1^*), \\
\frac{\partial^2 \tau_N}{\partial x \partial x_3} &= \frac{\partial}{\partial x}\left[(d_2, d_0^*, 1, 2, \ldots, N, N^*, \ldots, 2^*, 1^*)\right. \\
&\quad - (d_1, d_1^*, 1, 2, \ldots, N, N^*, \ldots, 2^*, 1^*) \\
&\quad \left. + (d_0, d_2^*, 1, 2, \ldots, N, N^*, \ldots, 2^*, 1^*)\right]
\end{aligned}$$

$$
\begin{aligned}
&= (d_3, d_0^*, 1, 2, \dots, N, N^*, \dots, 2^*, 1^*) \\
&\quad - (d_0, d_0^*, d_1, d_1^*, 1, 2, \dots, N, N^*, \dots, 2^*, 1^*) \\
&\quad + (d_0, d_3^*, 1, 2, \dots, N, N^*, \dots, 2^*, 1^*), \\
\frac{\partial^2 \tau_N}{\partial x_2^2} &= \frac{\partial}{\partial x_2} \big[-(d_1, d_0^*, 1, 2, \dots, N, N^*, \dots, 2^*, 1^*) \\
&\quad + (d_0, d_1^*, 1, 2, \dots, N, N^*, \dots, 2^*, 1^*) \big] \\
&= (d_3, d_0^*, 1, 2, \dots, N, N^*, \dots, 2^*, 1^*) \\
&\quad - (d_2, d_1^*, 1, 2, \dots, N, N^*, \dots, 2^*, 1^*) \\
&\quad - 2(d_0, d_0^*, d_1, d_1^*, 1, 2, \dots, N, N^*, \dots, 2^*, 1^*) \\
&\quad - (d_1, d_2^*, 1, 2, \dots, N, N^*, \dots, 2^*, 1^*) \\
&\qquad + (d_0, d_3^*, 1, 2, \dots, N, N^*, \dots, 2^*, 1^*).
\end{aligned}
$$

From the above calculations we obtain, for $\tau = \tau_N$,

$$
\begin{aligned}
\tau &= (\bullet), \\
\tau_x &= (d_0, d_0^*, \bullet), \\
\tau_{xx} - \tau_{x_2} &= -2(d_0^*, d_1, \bullet), \\
\tau_{xx} + \tau_{x_2} &= 2(d_0, d_1^*, \bullet), \\
\tau_{xxx} - \tau_{x_3} &= 3(d_1, d_1^*, \bullet), \\
\tau_{xxxx} - 4\tau_{x_3 x} + 3\tau_{x_2 x_2} &= 12(d_0, d_0^*, d_1, d_1^*, \bullet),
\end{aligned}
\tag{3.48}
$$

where we have used the abbreviated notation $\bullet$ for the list of indices $1, 2, \dots, N, N^*, \dots, 2^*, 1^*$ common to each pfaffian.

Let us rewrite the above results in terms of Maya diagrams:

	d_0	d_0^*	d_1	d_1^*	
$\tau =$					,
$\tau_x =$	○	○			,
$\tau_{xx} - \tau_{x_2} = -2$		○	○		,
$\tau_{xx} + \tau_{x_2} = \ \ 2$	○			○	,
$\tau_{xxx} - \tau_{x_3} = \ \ 3$			○	○	,
$\tau_{xxxx} - 4\tau_{x_3 x} + 3\tau_{x_2 x_2} = 12$	○	○	○	○	.

Therefore, the bilinear KP equation,

$$
\begin{aligned}
(D_1^4 - 4D_1D_3 + 3D_2^2)\tau \cdot \tau
&= 2[(\tau_{xxxx} - 4\tau_{x_3,x} + 3\tau_{x_2,x_2})\tau - 4(\tau_{xxx} - \tau_{x_3})\tau_x \\
&\quad + 3(\tau_{xx} - \tau_{x_2})(\tau_{xx} + \tau_{x_2})] \\
&= 0,
\end{aligned} \tag{3.49}
$$

is expressed by means of Maya diagrams as

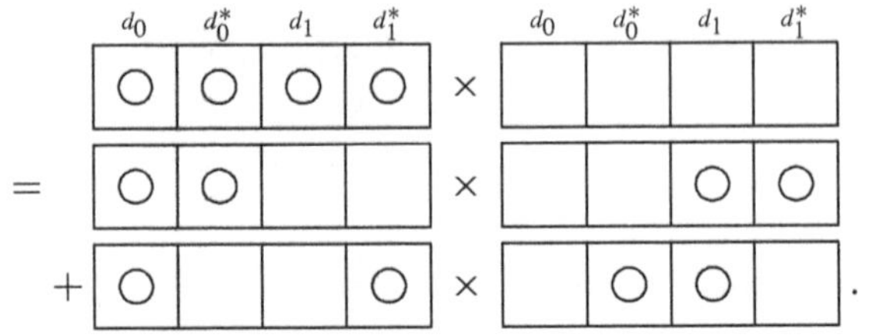

This is nothing but the Jacobi identity for determinants and also a special case of the pfaffian identity

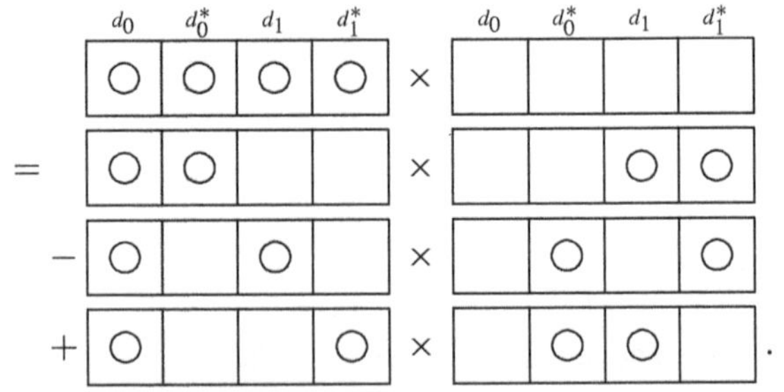

Hence, $\tau = \tau_N$ is a solution to the KP equation.

Remark

The second term on the right-hand side of the above equation is

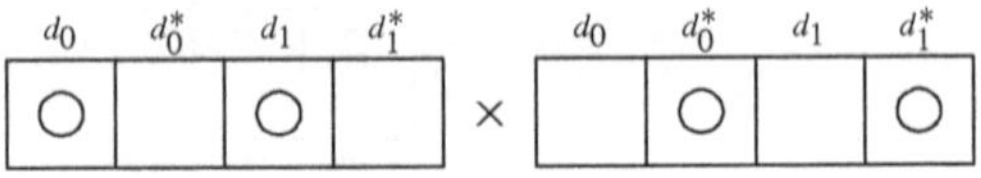

$$
\begin{aligned}
&= (d_0, d_1, 1, 2, \ldots, N, N^*, \ldots, 2^*, 1^*) \\
&\quad \times (d_0^*, d_1^*, 1, 2, \ldots, N, N^*, \ldots, 2^*, 1^*).
\end{aligned}
$$

However, the number of indices with and without superscript * in each pfaffian are not equal. Therefore, by virtue of the relations

$$(i, j) = (i^*, j^*) = 0, \qquad (d_n, i) = (d_m^*, j^*) = 0, \tag{3.50}$$

each of these pfaffians is zero. □

3.3 The BKP equation: pfaffian solutions

Among equations shown to possess N-soliton solutions in Section 1.8, the following pairs of equations have the same nonlinearity and dispersion. The only differences are in the coefficients of their nonlinear terms.

- Lax's fifth-order KdV equation (1.252)

$$u_t + 10\left(u^3 + \tfrac{1}{2}u_x^2 + uu_x\right)_x + u_{xxxxx} = 0, \tag{3.51a}$$

$$u = 2(\log f)_{xx}, \tag{3.51b}$$

$$\left[D_x(D_t + D_x^5) - \tfrac{5}{3}D_s(D_s + D_x^3)\right]f \cdot f = 0, \tag{3.51c}$$

where f satisfies simultaneously the bilinear equation involving auxiliary variable s

$$D_x(D_s + D_x^3)f \cdot f = 0. \tag{3.52}$$

- Sawada–Kotera equation (1.254)

$$u_t + 15(u^3 + uu_{xx})_x + u_{xxxxx} = 0, \tag{3.53a}$$

$$u = 2(\log f)_{xx}, \tag{3.53b}$$

$$D_x(D_t + D_x^5)f \cdot f = 0. \tag{3.53c}$$

- Model equations for shallow water waves

(i)

$$u_t - u_{xxt} - 4uu_t + 2u_x \int_x^\infty u_t \, dx' + u_x = 0, \tag{3.54a}$$

$$u = 2(\log f)_{xx}, \tag{3.54b}$$

$$\left[D_x(D_t - D_tD_x^2 + D_x) + \tfrac{1}{3}D_t(D_s + D_x^3)\right]f \cdot f = 0, \tag{3.54c}$$

where f also satisfies

$$D_x(D_s + D_x^3)f \cdot f = 0; \tag{3.55}$$

(ii)
$$u_t - u_{xxt} - 3uu_t + 3u_x \int_x^{\infty} u_t \, dx' + u_x = 0, \tag{3.56a}$$
$$u = 2(\log f)_{xx}, \tag{3.56b}$$
$$D_x(D_t - D_t D_x^2 + D_x) f \cdot f = 0. \tag{3.56c}$$

Soliton solutions for (3.51) and (3.54) were discovered by an inverse scattering method [45]. To the best of the author's knowledge, however, those for (3.53) and (3.56) were not found by such a method. Rather, they were found by the direct method. In fact, (3.53) and (3.56) belong to the class of BKP-type equations, and the most suitable expressions for their soliton solutions are pfaffians [31].

Remarks

(1) The KP equation is associated with an A-type group and the BKP equation with a B-type group. The BKP equation was given this name because it is a B-type KP equation.
(2) In order to find soliton solutions for BKP-type equations by the inverse method, it is necessary to change the structure of the Gel'fand–Levitan integral equation because the solutions are expressed not as determinants but as pfaffians [46]. □

In this section, we will express the soliton solutions of the BKP equation in terms of a pfaffian, and we will show that the bilinear BKP equation is equivalent to a pfaffian identity. The class of BKP-type equations (BKP hierarchy) includes

$$\left[(D_3 - D_1^3)D_{-1} + 3D_1^2\right]\tau \cdot \tau = 0, \tag{3.57a}$$
$$(D_1^6 - 5D_1^3 D_3 - 5D_3^2 + 9D_1 D_5)\tau \cdot \tau = 0, \tag{3.57b}$$
$$\cdots$$

Remark

The first of this class, (3.57a), is transformed through the dependent variable transformation,

$$w = 2(\log \tau)_x, \tag{3.58}$$

into the nonlinear partial differential equation

$$w_{yt} - w_{xxxy} - 3(w_x w_y)_x + 3w_{xx} = 0, \tag{3.59}$$

where we have put $x_1 = x$, $x_{-1} = y$, $x_3 = t$. □

In order to investigate the solution of (3.57a), let us consider its two-soliton expression

$$\tau_2 = 1 + \exp(\eta_1) + \exp(\eta_2) + b_{12}\exp(\eta_1 + \eta_2). \qquad (3.60)$$

In a similar way to the KP equation, if we introduce parameters p_i and q_i we may rewrite $\exp(\eta_i)$ as

$$\begin{aligned}\exp(\eta_i) &= \exp(\xi_i + \widehat{\xi}_i),\\ \xi_i &= p_i^{-1}x_{-1} + p_i x_1 + p_i^3 x_3 + p_i^5 x_5 + \xi_i^0,\\ \widehat{\xi}_i &= q_i^{-1}x_{-1} + q_i x_1 + q_i^3 x_3 + q_i^5 x_5 + \widehat{\xi}_i^0,\end{aligned} \qquad (3.61)$$

and the dispersion relation is automatically satisfied. Also, the phase shift term b_{12} is given by

$$b_{12} = \frac{(p_1 - p_2)(p_1 - q_2)(q_1 - p_2)(q_1 - q_2)}{(p_1 + p_2)(p_1 + q_2)(q_1 + p_2)(q_1 + q_2)}. \qquad (3.62)$$

If we put $b_{ij} = \exp(B_{ij})$, the N-soliton solution is expressed as

$$\tau_N = \sum \exp\left[\sum_{i=1}^{N} \mu_i \eta_i + \sum_{i<j}^{(N)} B_{ij}\mu_i\mu_j\right], \qquad (3.63)$$

where $\sum$ denotes the summation over all possible combinations of $\mu_1 = 0, 1,\ \mu_2 = 0, 1, \dots,\ \mu_N = 0, 1$, and $\sum_{i<j}^{(N)}$ is the sum over all pairs i, j $(i < j)$ chosen from $\{1, 2, \dots, N\}$.

The N-soliton solution τ_N can also be expressed as the Nth-order pfaffian,

$$\tau_N = (1, 2, 3, \dots, 2N), \qquad (3.64)$$

where $(i, j) = c_{ij} + \int^x D_x f_i(x) \cdot f_j(x)\,\mathrm{d}x \quad (x = x_1)$ and $f_i(x)$ for $i = 1, 2, 3, \dots$ satisfy the linear differential equations

$$\frac{\partial}{\partial x_n} f_i(x) = \frac{\partial^n}{\partial x^n} f_i(x) \quad (n = -1, 1, 3, 5, \dots). \qquad (3.65)$$

For $n = -1$ this means that

$$\frac{\partial}{\partial x_{-1}} f_i(x) = \int^x f_i(x)\,\mathrm{d}x. \qquad (3.66)$$

We also note that $c_{ij} = -c_{ji}$.

Remarks

(1) An N-soliton solution is a solution with $3N$ parameters p_i, q_i, ξ_i^0 for $i = 1, 2, \dots, N$. Since τ_N is expressed as a pfaffian containing functions $f_i(x)$

with arbitrary parameters, it represents other solutions as well as soliton solutions.

(2) We choose a lower limit of the above integral to be $x = \pm\infty$ (the value x such that the value of an integrand is equal to zero), but this is not an essential restriction. The result is the same for any choice of the lower limit.

(3) Using integration by parts, each pfaffian entry (i, j) is

$$\begin{aligned}(i, j) &= c_{ij} + \int^x D_x f_i \cdot f_j \, \mathrm{d}x \\ &= c_{ij} + 2\int^x \frac{\partial f_i}{\partial x} f_j \, \mathrm{d}x - f_i f_j. \end{aligned} \tag{3.67}$$

Therefore, from the relation (2.92) between a pfaffian and a determinant, the square of the N-soliton solution τ_N can be written as the determinant

$$\tau_N^2 = \left| c_{ij} + 2\int^x \frac{\partial f_i}{\partial x} f_j \, \mathrm{d}x \right|_{1\le i,j\le 2N}. \tag{3.68}$$

This determinant is nothing but the grammian solution of the KP equation, τ_{KP}. Hence, we have

$$\tau_{\mathrm{KP}} = \tau_{\mathrm{BKP}}^2. \tag{3.69}$$

(4) By choosing $c_{12} = c_{34} = 1$, $c_{13} = c_{14} = c_{23} = c_{24} = 0$, $f_1 = \exp \xi_1$, $f_2 = \exp \widehat{\xi_1}$, $f_3 = \exp \xi_2$ and $f_4 = \exp \widehat{\xi_2}$, the two-soliton solution is

$$\begin{aligned}\tau_2 &= (1,2)(3,4) - (1,3)(2,4) + (1,4)(2,3) \\ &= \left[1 + \frac{p_1 - q_1}{p_1 + q_1} \exp(\xi_1 + \widehat{\xi_1})\right]\left[1 + \frac{p_2 - q_2}{p_2 + q_2} \exp(\xi_2 + \widehat{\xi_2})\right] \\ &\quad - \frac{p_1 - p_2}{p_1 + p_2} \exp(\xi_1 + \xi_2) \times \frac{q_1 - q_2}{q_1 + q_2} \exp(\widehat{\xi_1} + \widehat{\xi_2}) \\ &\quad + \frac{p_1 - q_2}{p_1 + q_2} \exp(\xi_1 + \widehat{\xi_2}) \times \frac{q_1 - p_2}{q_1 + p_2} \exp(\widehat{\xi_1} + \xi_2). \end{aligned} \tag{3.70}$$

Putting

$$\widehat{\eta_i} = \xi_i + \widehat{\xi_i} + \delta_i, \text{ where } \exp \delta_i = \frac{p_i - q_i}{p_i + q_i}, \tag{3.71}$$

we may rewrite τ_2 as

$$\tau_2 = 1 + \exp(\widehat{\eta_1}) + \exp(\widehat{\eta_2}) + b_{12} \exp(\widehat{\eta_1} + \widehat{\eta_2}), \tag{3.72}$$

which coincides with the two-soliton solution found by the perturbation method. □

Let us confirm that the pfaffian expression for τ satisfies the BKP-type equation (3.57a),

$$\left[(D_3 - D_1^3)D_{-1} + 3D_1^2\right]\tau \cdot \tau = 0. \tag{3.73}$$

This is expressed in terms of normal derivatives as

$$\begin{aligned}&\left[\frac{\partial}{\partial x_{-1}}\left(\frac{\partial}{\partial x_3} - \frac{\partial^3}{\partial x^3}\right)\tau + 3\frac{\partial^2}{\partial x^2}\tau\right]\tau + 3\left[\left(\frac{\partial^3}{\partial x_{-1}\partial x^2} - \frac{\partial}{\partial x}\right)\tau\right]\frac{\partial \tau}{\partial x}\\ &\quad - 3\left[\frac{\partial^2}{\partial x_{-1}\partial x}\tau\right]\frac{\partial^2 \tau}{\partial x^2} - \left[\left(\frac{\partial}{\partial x_3} - \frac{\partial^3}{\partial x^3}\right)\tau\right]\frac{\partial \tau}{\partial x_{-1}} = 0. \end{aligned}\tag{3.74}$$

From the formula

$$(i, j) = c_{ij} + \int^x D_x f_i(x) \cdot f_j(x)\, \mathrm{d}x, \tag{3.75}$$

we obtain

$$\begin{aligned}\frac{\partial}{\partial x}(i, j) &= \left[\frac{\partial}{\partial x} f_i\right] f_j - f_i\left[\frac{\partial}{\partial x} f_j\right]\\ &= (d_0, d_1, i, j),\\ \frac{\partial}{\partial x_{-1}}(i, j) &= \int^x \left[\frac{\partial^2 f_i}{\partial x_{-1}\partial x} f_j + \frac{\partial f_i}{\partial x}\frac{\partial f_j}{\partial x_{-1}} - \frac{\partial f_i}{\partial x_{-1}}\frac{\partial f_j}{\partial x} - f_i \frac{\partial^2 f_j}{\partial x_{-1}\partial x}\right]\mathrm{d}x\\ &= f_i\left[\frac{\partial}{\partial x_{-1}} f_j\right] - \left[\frac{\partial}{\partial x_{-1}} f_i\right] f_j\\ &= (d_{-1}, d_0, i, j),\\ \frac{\partial}{\partial x_3}(i, j) &= \int^x \left[\frac{\partial^2 f_i}{\partial x_3 \partial x} f_j + \frac{\partial f_i}{\partial x}\frac{\partial f_j}{\partial x_3} - \frac{\partial f_i}{\partial x_3}\frac{\partial f_j}{\partial x} - f_i \frac{\partial^2 f_j}{\partial x_3 \partial x}\right]\mathrm{d}x\\ &= \frac{\partial^3 f_i}{\partial x^3} f_j - f_i \frac{\partial^3 f_j}{\partial x^3} - 2\left[\frac{\partial^2 f_i}{\partial x^2}\frac{\partial f_j}{\partial x} - \frac{\partial f_i}{\partial x}\frac{\partial^2 f_j}{\partial x^2}\right]\\ &= (d_0, d_3, i, j) - 2(d_1, d_2, i, j),\end{aligned}$$

where we define

$$(d_n, i) = \frac{\partial^n}{\partial x^n} f_i(x) \quad (n = -1, 0, 1, 2, \ldots). \tag{3.76}$$

Consequently, the derivatives of τ_N,

$$\tau_N = (1, 2, 3, \ldots, 2N) = (\bullet), \tag{3.77}$$

are given by

$$\frac{\partial}{\partial x}\tau_N = (d_0, d_1, \bullet), \tag{3.78a}$$

$$\frac{\partial^2}{\partial x^2}\tau_N = (d_0, d_2, \bullet), \tag{3.78b}$$

$$\frac{\partial^3}{\partial x^3}\tau_N = (d_1, d_2, \bullet) + (d_0, d_3, \bullet), \tag{3.78c}$$

$$\frac{\partial^3}{\partial x_{-1}\partial x^2}\tau_N = (d_{-1}, d_2, \bullet) + (d_0, d_1, \bullet), \tag{3.78d}$$

$$\frac{\partial}{\partial x_{-1}}\tau_N = (d_{-1}, d_0, \bullet), \tag{3.78e}$$

$$\frac{\partial^2}{\partial x_{-1}\partial x}\tau_N = (d_{-1}, d_1, \bullet), \tag{3.78f}$$

$$\frac{\partial}{\partial x_3}\tau_N = (d_0, d_3, \bullet) - 2(d_1, d_2, \bullet), \tag{3.78g}$$

$$\frac{\partial}{\partial x_{-1}}\left(\frac{\partial}{\partial x_3} - \frac{\partial^3}{\partial x^3}\right)\tau_N = -3\left[(d_0, d_2, \bullet) + (d_{-1}, d_0, d_1, d_2, \bullet)\right]. \tag{3.78h}$$

Substituting the above results into (3.73), we obtain

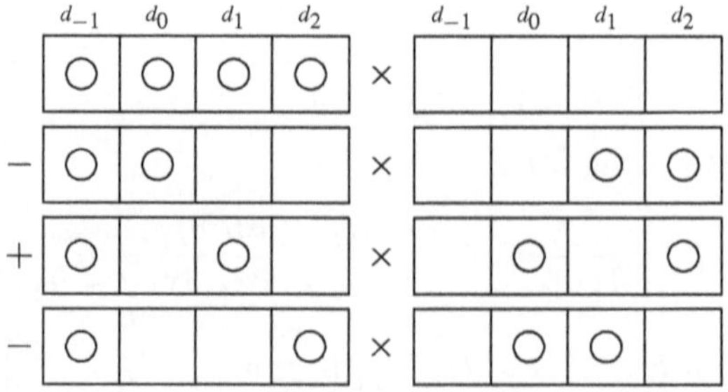

$= 0,$

which is nothing but a pfaffian identity. Therefore, τ_N solves the BKP-type equation (3.57a).

Through similiar calculations that are a little more complicated, we can confirm that τ_N also satisfies (3.57b).

3.4 The coupled KP equation

3.4.1 Wronski-type pfaffian solutions

In Section 3.2 we showed two things about the bilinear KP equation

$$(D_1^4 - 4D_1D_3 + 3D_2^2)\tau \cdot \tau = 0. \tag{3.79}$$

First, if τ is expressed as a wronskian determinant, it is equivalent to the Plücker relation,

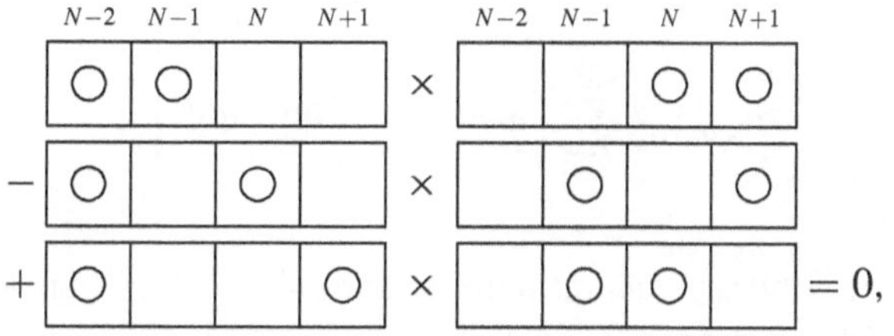

and, secondly, if τ is expressed as a grammian determinant, it is equivalent to the Jacobi identity

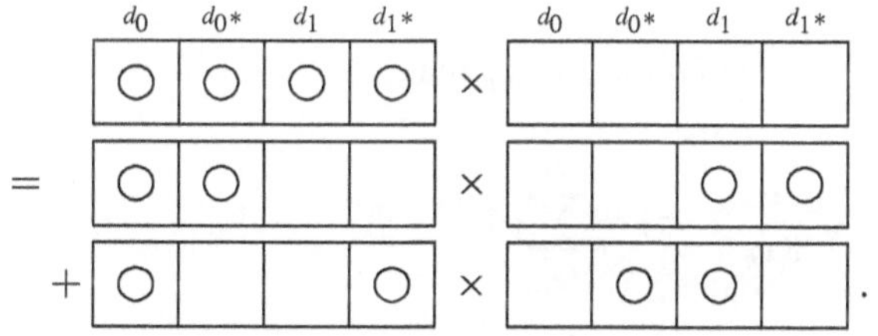

Both the Plücker relation and the Jacobi identity are special cases of the pfaffian identity

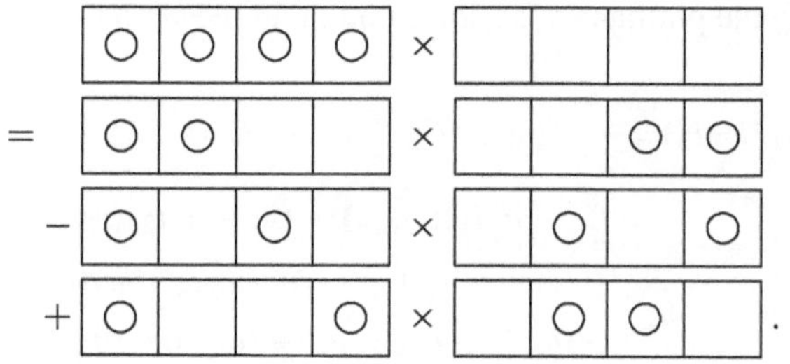

Since a determinant is a special case of a pfaffian, it may be possible to obtain new soliton equations, including the KP equation as a special case, by extending the expression of the τ-function as a determinant to a more general pfaffian. Let us pursue this possibility.

First, consider a pfaffian with entries (l, m) satisfying the differential rules with respect to the variables $x_1, x_2, \ldots$:

$$\frac{\partial}{\partial x_n}(l, m) = (l + n, m) + (l, m + n). \tag{3.80}$$

We call this a *Wronski-type pfaffian.*

Various examples of Wronski-type pfaffians may be considered. Among them, we consider one with the entries

$$(l, m) \equiv \sum_{k=1}^{M} \left[\Phi_k^{(l)} \Psi_k^{(m)} - \Phi_k^{(m)} \Psi_k^{(l)} \right], \tag{3.81}$$

where M is an arbitrary natural number and $\Phi_k^{(l)}$ and $\Psi_k^{(l)}$ stand for the lth derivatives with respect to x $(= x_1)$ of functions Φ_k and Ψ_k satisfying

$$\frac{\partial}{\partial x_n}\Phi_k = \Phi_k^{(n)}, \quad \frac{\partial}{\partial x_n}\Psi_k = \Psi_k^{(n)}. \tag{3.82}$$

Under the above assumption, we see that

$$\begin{aligned} \frac{\partial}{\partial x_n}(l, m) &= \sum_{k=1}^{M} [\Phi_k^{(l+n)} \Psi_k^{(m)} - \Phi_k^{(m+n)} \Psi_k^{(l)} \\ &\quad + \Phi_k^{(l)} \Psi_k^{(m+n)} - \Phi_k^{(m)} \Psi_k^{(l+n)}] \\ &= (l + n, m) + (l, m + n). \end{aligned} \tag{3.83}$$

As we have done previously, we will investigate the differential rules for this pfaffian using the pfaffian expansion formula. In the simplest case, we have

$$\begin{aligned} \frac{\partial}{\partial x_n}(i_0, i_1, i_2, i_3) &= \frac{\partial}{\partial x_n}[(i_0, i_1)(i_2, i_3) - (i_0, i_2)(i_1, i_3) + (i_0, i_3)(i_1, i_2)] \\ &= (i_0 + n, i_1)(i_2, i_3) - (i_0 + n, i_2)(i_1, i_3) \\ &\quad + (i_0 + n, i_3)(i_1, i_2) + (i_0, i_1 + n)(i_2, i_3) \\ &\quad - (i_0, i_2 + n)(i_1, i_3) + (i_0, i_3 + n)(i_1, i_2) \end{aligned}$$

$$
\begin{aligned}
&+ (i_0, i_1)(i_2 + n, i_3) - (i_0, i_2)(i_1 + n, i_3)\\
&+ (i_0, i_3)(i_1 + n, i_2) + (i_0, i_1)(i_2, i_3 + n)\\
&- (i_0, i_2)(i_1, i_3 + n) + (i_0, i_3)(i_1, i_2 + n)\\
&= (i_0 + n, i_1, i_2, i_3) + (i_0, i_1 + n, i_2, i_3)\\
&+ (i_0, i_1, i_2 + n, i_3) + (i_0, i_1, i_2, i_3 + n).
\end{aligned}
$$

From this, we see that the differential rule for the Wronski-type pfaffian $\tau^{\mathrm{W}} = (i_0, i_1, \ldots, i_{2N-1})$, where the superscript W stands for wronskian, is

$$
\frac{\partial}{\partial x_n}(i_0, i_1, \ldots, i_{2N-1}) = \sum_{k=0}^{2N-1} (i_0, i_1, \ldots, i_k + n, \ldots, i_{2N-1}). \tag{3.84}
$$

On the other hand, if we introduce a wronskian defined by

$$
|r(0), r(1), r(2), \ldots, r(2N-1)| \equiv \begin{vmatrix} r_1(0) & r_1(1) & \ldots & r_1(2N-1) \\ r_2(0) & r_2(1) & \ldots & r_2(2N-1) \\ \vdots & \vdots & & \vdots \\ r_{2N}(0) & r_{2N}(1) & \ldots & r_{2N}(2N-1) \end{vmatrix}, \tag{3.85}
$$

where $r_j(n)$ is given by

$$
r_j(n) \equiv \frac{\partial^n}{\partial x^n} r_j(0), \qquad \frac{\partial^n}{\partial x^n} r_j(m) = r_j(m+n), \tag{3.86}
$$

then the differential rule is, as is well known, given by

$$
\begin{aligned}
&\frac{\partial}{\partial x_n}|r(i_0), r(i_1), \ldots, r(i_{2N-1})|\\
&\quad = \sum_{k=0}^{2N-1} |r(i_0), r(i_1), \ldots, r(i_k + n), \ldots, r(i_{2N-1})|.
\end{aligned} \tag{3.87}
$$

This has exactly the same form as the differential rule for the pfaffian τ^{W}. This is the reason that τ^{W} is called a Wronski-type pfaffian.

Substituting the pfaffian $\tau^{\mathrm{W}} = (0, 1, \ldots, 2N-1)$ into the left-hand side of the KP equation, we have

$$(D_1^4 - 4D_1D_3 + 3D_2^2)\tau^{\mathrm{W}} \cdot \tau^{\mathrm{W}} = 24$$

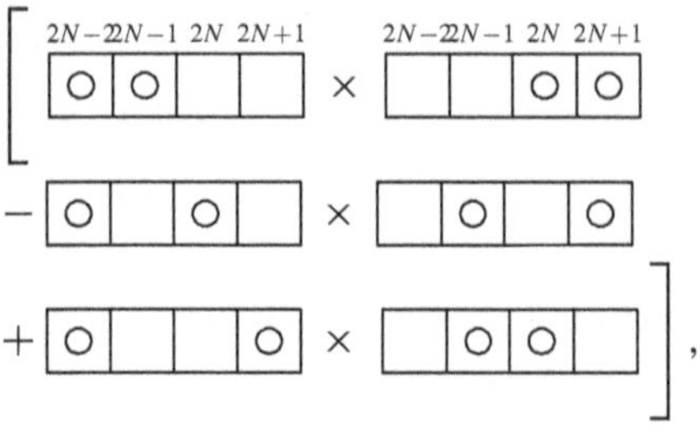

through the same calculation performed in the case of the wronskian solution of the KP equation. By employing a pfaffian identity, the right-hand side of the above equation is written as

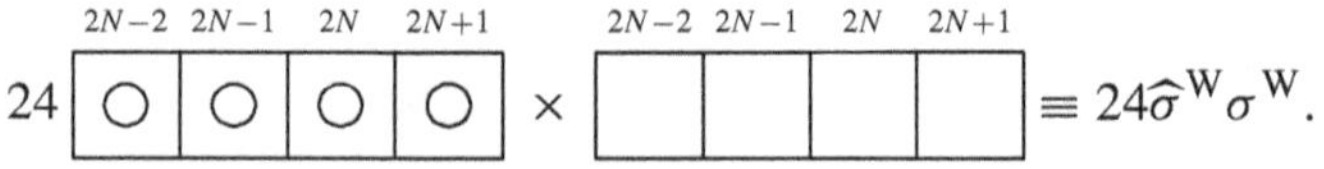

We remark that we have $\sigma^{\mathrm{W}} = \widehat{\sigma}^{\mathrm{W}} = 0$ if τ is a wronskian.

The pfaffians $\sigma^{\mathrm{W}}, \widehat{\sigma}^{\mathrm{W}}$ introduced here,

$$\begin{aligned} \sigma^{\mathrm{W}} &= (0, 1, 2, \ldots, 2N-3), \\ \widehat{\sigma}^{\mathrm{W}} &= (0, 1, 2, \ldots, 2N, 2N+1), \end{aligned} \tag{3.88}$$

have order which are two less and two more, respectively, than that of τ. Taking this fact into account and searching for a pfaffian identity involving σ^{W} and τ, we obtain the Maya diagram expression

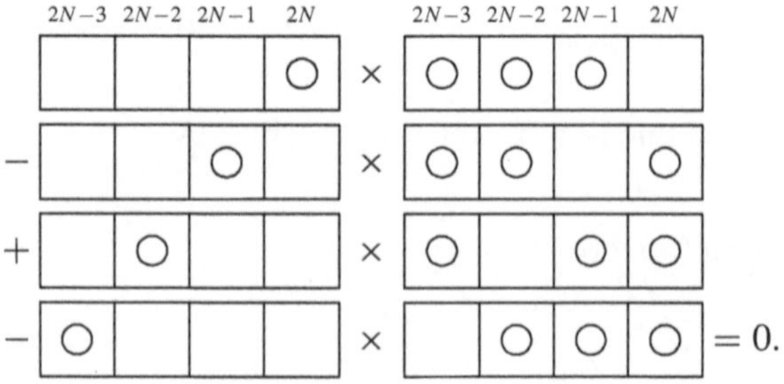

This identity is equivalent to the bilinear equation

$$(D_1^3 + 2D_3 + 3D_1D_2)\sigma^{\mathrm{W}} \cdot \tau^{\mathrm{W}} = 0. \tag{3.89}$$

Following the same procedure, we obtain a bilinear equation involving $\widehat{\sigma}^{\mathrm{W}}$ and τ^{W}:

$$(D_1^3 + 2D_3 - 3D_1D_2)\widehat{\sigma}^{\mathrm{W}} \cdot \tau^{\mathrm{W}} = 0. \tag{3.90}$$

Remark

These two bilinear equations have the same form as the second modified KP equation introduced by Jimbo and Miwa. However, σ^{W} and τ^{W} in the second modified KP equation are expressed in determinantal form, not in pfaffian form. □

Let us summarize the above results. If we define pfaffians $\tau^{\mathrm{W}}, \sigma^{\mathrm{W}}, \widehat{\sigma}^{\mathrm{W}}$ by

$$\begin{aligned} \tau^{\mathrm{W}} &\equiv (0, 1, \ldots, 2N-1), \\ \sigma^{\mathrm{W}} &\equiv (0, 1, 2, \ldots, 2N-3), \\ \widehat{\sigma}^{\mathrm{W}} &\equiv (0, 1, 2, \ldots, 2N, 2N+1), \end{aligned} \tag{3.91}$$

where

$$\frac{\partial}{\partial x_n}(l, m) = (l+n, m) + (l, m+n),$$

we have the coupled bilinear equations,

$$\begin{aligned} (D_1^4 - 4D_1D_3 + 3D_2^2)\tau^{\mathrm{W}} \cdot \tau^{\mathrm{W}} &= 24\widehat{\sigma}^{\mathrm{W}}\sigma^{\mathrm{W}}, \\ (D_1^3 + 2D_3 + 3D_1D_2)\sigma^{\mathrm{W}} \cdot \tau^{\mathrm{W}} &= 0, \\ (D_1^3 + 2D_3 - 3D_1D_2)\widehat{\sigma}^{\mathrm{W}} \cdot \tau^{\mathrm{W}} &= 0. \end{aligned} \tag{3.92}$$

Through the dependent variable transformations

$$u = 2(\log \tau^{\mathrm{W}})_{xx}, \quad v = \sigma^{\mathrm{W}}/\tau^{\mathrm{W}}, \quad \widehat{v} = \widehat{\sigma}^{\mathrm{W}}/\tau^{\mathrm{W}}, \tag{3.93}$$

the bilinear equations are equivalent to the coupled nonlinear partial differential equations

$$\begin{aligned} (4u_t - 6uu_x - u_{xxx})_x - 3u_{yy} + 24(v\widehat{v})_{xx} &= 0, \\ 2v_t + 3uv_x + v_{xxx} + 3\left(v_{xy} + v\int^x u_y \,\mathrm{d}x\right) &= 0, \\ 2\widehat{v}_t + 3u\widehat{v}_x + \widehat{v}_{xxx} - 3\left(\widehat{v}_{xy} + \widehat{v}\int^x u_y \,\mathrm{d}x\right) &= 0, \end{aligned} \tag{3.94}$$

where $x = x_1,\ y = x_2,\ t = x_3$, which we call the *coupled KP equation*.

Remarks

(1) Putting $t = 2T$, $v = \widehat{v} = (-b/12)^{1/2}\phi$ and omitting the y-dependence in the above system, we have

$$\begin{aligned} u_T - \frac{1}{2}(u_{xxx} + 6uu_x) &= 2b\phi\phi_x, \\ \phi_T + \phi_{xxx} + 3u\phi_x &= 0. \end{aligned} \tag{3.95}$$

This system is called the *coupled KdV equation* [47] and is a specialized version of a 4-reduction of the KP hierarchy. In fact it is associated with the Kač–Moody algebra $C_2^{(4)}$ [15]. Therefore, the system (3.94) may also be thought of as a two-dimensional version of the coupled KdV equation. This is why we called (3.94) the coupled KP equation.

(2) The group acting on the coupled KP equation is completely unknown[1] (the author believes that it should also be associated with a Kač–Moody algebra). From the standpoint of group theory, the A-type group acting on the KP equation is the most general, and we consider that the group acting on the BKP equation is a specialization. However, this idea is not consistent with the author's viewpoint that pfaffians are more general than determinants. Hence, we have avoided a discussion of soliton equations from a group theoretical viewpoint.

(3) For arbitrary natural numbers M, N, let us consider the pfaffians

$$\begin{aligned} \tau^{\mathrm{W}} &\equiv (b_1, b_2, \ldots, b_M, 0, 1, \ldots, N-1), \\ \sigma^{\mathrm{W}} &\equiv (b_1, b_2, \ldots, b_M, 0, 1, \ldots, N-3), \\ \widehat{\sigma}^{\mathrm{W}} &\equiv (b_1, b_2, \ldots, b_M, 0, 1, \ldots, N+1), \end{aligned} \tag{3.96}$$

where

$$\frac{\partial}{\partial x_n}(l, m) = (l+n, m) + (l, m+n),$$

$$(b_i, b_j) = 0, \quad (b_i, l) = \frac{\partial^l}{\partial x^l}\phi_i,$$

$$\frac{\partial}{\partial x_n}\phi_i = \frac{\partial^n}{\partial x^n}\phi_i,$$

for $l, m = 0, 1, \ldots, N+1$, $i, j = 1, 2, \ldots, M$ and $n = 1, 2, 3, \ldots$. In the case $M = N$, τ^{W} gives an nth-order wronskian and therefore a solution of the KP equation. Otherwise, if $M > N$, we have $\tau^{\mathrm{W}} = 0$, and if $M < N$,

[1] Translators' note: It has since been realized that the coupled KP equation is associated with the affine Lie algebra D_∞. See ref. [9], p. 976.

τ^{W} is a hybrid solution in which solutions of the KP equation and of the BKP equation coexist. □

3.4.2 Gramm-type pfaffian solutions

We have seen that the KP equation has solutions expressed in wronskian and in grammian form. In a simliar way, the coupled KP equation,

$$(D_1^4 - 4D_1D_3 + 3D_2^2)\tau \cdot \tau = 24\widehat{\sigma}\sigma, \tag{3.97a}$$

$$(D_1^3 + 2D_3 + 3D_1D_2)\sigma \cdot \tau = 0, \tag{3.97b}$$

$$(D_1^3 + 2D_3 - 3D_1D_2)\widehat{\sigma} \cdot \tau = 0, \tag{3.97c}$$

has solutions that can be expressed as Wronski-type pfaffians $\tau^{\mathrm{W}}, \sigma^{\mathrm{W}}, \widehat{\sigma}^{\mathrm{W}}$ and as Gramm-type pfaffians $\tau^{\mathrm{G}}, \sigma^{\mathrm{G}}, \widehat{\sigma}^{\mathrm{G}}$ [48]. The latter are

$$\begin{aligned} \tau^{\mathrm{G}} &\equiv (1, 2, \dots, 2N), \\ \sigma^{\mathrm{G}} &\equiv (c_1, c_0, 1, 2, \dots, 2N), \\ \widehat{\sigma}^{\mathrm{G}} &\equiv (d_0, d_1, 1, 2, \dots, 2N), \end{aligned} \tag{3.98}$$

where the different types of pfaffian entries are defined by

$$\begin{aligned} (i, j) &\equiv c_{ij} + \int^x (f_i g_j - f_j g_i)\,\mathrm{d}x, \quad c_{ij} = -c_{ji}, \\ (d_n, i) &\equiv \frac{\partial^n}{\partial x^n} f_i, \\ (c_n, i) &\equiv \frac{\partial^n}{\partial x^n} g_i, \\ (d_m, d_n) &= (c_m, c_n) = (c_m, d_n) = 0. \end{aligned} \tag{3.99}$$

In the above definition of (i, j), the lower limit of integration is chosen so that the functions f_i, g_j and their derivatives are zero. As in other similar cases, this is not an essential restriction; constants coming from the lower limit can be transferred into the term c_{ij}. We also see that f_i, g_i $(i = 1, 2, \dots, 2N)$ satisfy the differential equations

$$\begin{aligned} \frac{\partial}{\partial x_n} f_i &= \frac{\partial^n}{\partial x^n} f_i, \\ \frac{\partial}{\partial x_n} g_i &= (-1)^{n-1} \frac{\partial^n}{\partial x^n} g_i \quad (n = 1, 2, 3, \dots). \end{aligned} \tag{3.100}$$

Using the formulae

$$
\begin{aligned}
\frac{\partial}{\partial x}(i,j) &= f_i g_j - f_j g_i \\
&= (c_0, d_0, i, j), \\
\frac{\partial}{\partial x_2}(i,j) &= \frac{\partial f_i}{\partial x} g_j - f_i \frac{\partial g_j}{\partial x} - \frac{\partial f_j}{\partial x} g_i + f_j \frac{\partial g_i}{\partial x} \\
&= (c_0, d_1, i, j) - (c_1, d_0, i, j), \\
\frac{\partial}{\partial x_3}(i,j) &= \frac{\partial^2 f_i}{\partial x^2} g_j - \frac{\partial^2 f_j}{\partial x^2} g_i - \frac{\partial f_i}{\partial x}\frac{\partial g_j}{\partial x} + \frac{\partial f_j}{\partial x}\frac{\partial g_i}{\partial x} \\
&\quad + f_i \frac{\partial^2 g_j}{\partial x^2} - f_j \frac{\partial^2 g_i}{\partial x^2} \\
&= (c_0, d_2, i, j) - (c_1, d_1, i, j) + (c_2, d_0, i, j),
\end{aligned}
$$

and the same procedures as for the KP equation, we obtain expressions for the derivatives of τ^{G} and σ^{G}:

$$
\begin{array}{rl}
 & \begin{array}{cccc} c_0 & d_0 & c_1 & d_1 \end{array} \\
\tau^{\mathrm{G}} = & \begin{array}{|c|c|c|c|}\hline & & & \\ \hline\end{array}\,, \\
\tau^{\mathrm{G}}_x = & \begin{array}{|c|c|c|c|}\hline \bigcirc & \bigcirc & & \\ \hline\end{array}\,, \\
\tau^{\mathrm{G}}_{xx} - \tau^{\mathrm{G}}_{x_2} = -2 & \begin{array}{|c|c|c|c|}\hline & \bigcirc & \bigcirc & \\ \hline\end{array}\,, \\
\tau^{\mathrm{G}}_{xx} + \tau^{\mathrm{G}}_{x_2} = 2 & \begin{array}{|c|c|c|c|}\hline \bigcirc & & & \bigcirc \\ \hline\end{array}\,, \\
\tau^{\mathrm{G}}_{xxx} - \tau^{\mathrm{G}}_{x_3} = 3 & \begin{array}{|c|c|c|c|}\hline & & \bigcirc & \bigcirc \\ \hline\end{array}\,, \\
\tau^{\mathrm{G}}_{xxxx} - 4\tau^{\mathrm{G}}_{x_3 x} + 3\tau^{\mathrm{G}}_{x_2 x_2} = 12 & \begin{array}{|c|c|c|c|}\hline \bigcirc & \bigcirc & \bigcirc & \bigcirc \\ \hline\end{array}\,,
\end{array}
$$

$$
\begin{array}{rl}
 & \begin{array}{cccc} c_0 & c_1 & c_2 & d_0 \end{array} \\
\sigma^{\mathrm{G}} = - & \begin{array}{|c|c|c|c|}\hline \bigcirc & \bigcirc & & \\ \hline\end{array}\,, \\
\sigma^{\mathrm{G}}_x = - & \begin{array}{|c|c|c|c|}\hline \bigcirc & & \bigcirc & \\ \hline\end{array}\,, \\
\sigma^{\mathrm{G}}_{xx} + \sigma^{\mathrm{G}}_{x_2} = -2 & \begin{array}{|c|c|c|c|}\hline & \bigcirc & \bigcirc & \\ \hline\end{array}\,, \\
\sigma^{\mathrm{G}}_{xxx} + 2\sigma^{\mathrm{G}}_{x_3} + 3\sigma^{\mathrm{G}}_{x_2 x} = -6 & \begin{array}{|c|c|c|c|}\hline \bigcirc & \bigcirc & \bigcirc & \bigcirc \\ \hline\end{array}\,, \\
\tau^{\mathrm{G}}_{xxx} + 2\tau^{\mathrm{G}}_{x_3} + 3\tau^{\mathrm{G}}_{x_2 x} = 6 & \begin{array}{|c|c|c|c|}\hline & & \bigcirc & \bigcirc \\ \hline\end{array}\,.
\end{array}
$$

We have omitted the indices $\{1, 2, \ldots, 2N\}$ which are common to every pfaffian. Therefore, the coupled KP equation (3.97a,b),

$$\begin{aligned}
&(\tau^{\mathrm{G}}_{xxxx} - 4\tau^{\mathrm{G}}_{x_3x} + 3\tau^{\mathrm{G}}_{x_2x_2})\tau^{\mathrm{G}} - 4(\tau^{\mathrm{G}}_{xxx} - \tau^{\mathrm{G}}_{x_3})\tau^{\mathrm{G}}_x \\
&\quad + 3(\tau^{\mathrm{G}}_{xx} - \tau^{\mathrm{G}}_{x_2})(\tau^{\mathrm{G}}_{xx} + \tau^{\mathrm{G}}_{x_2}) = 12\sigma^{\mathrm{G}}\widehat{\sigma}^{G},
\end{aligned}$$

$$\begin{aligned}
&(\sigma^{\mathrm{G}}_{xxx} + 2\sigma^{\mathrm{G}}_{x_3} + 3\sigma^{\mathrm{G}}_{x_2x})\tau^{\mathrm{G}} - 3(\sigma^{\mathrm{G}}_{xx} + \sigma^{\mathrm{G}}_{x_2})\tau^{\mathrm{G}}_x \\
&\quad + 3\sigma^{\mathrm{G}}_x(\tau^{\mathrm{G}}_{xx} - \tau^{\mathrm{G}}_{x_2}) - \sigma^{\mathrm{G}}(\tau^{\mathrm{G}}_{xxx} + 2\tau^{\mathrm{G}}_{x_3} + 3\tau^{\mathrm{G}}_{x_2x}) = 0,
\end{aligned}$$

may be expressed in terms of Maya diagrams as

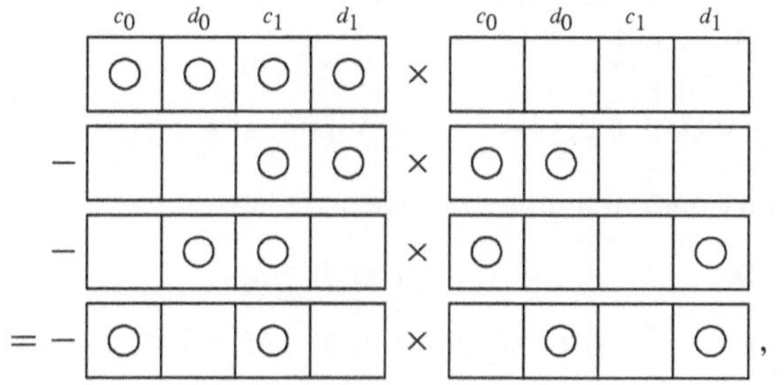

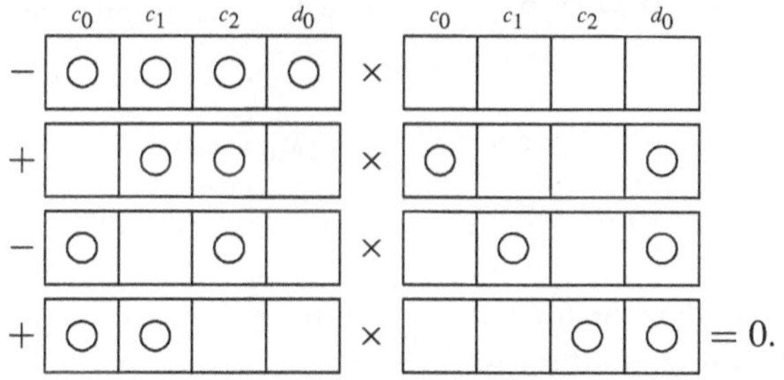

These are simply pfaffian identities. Equation (3.97c) is obtained by interchanging c and d in the above equation. Hence, a soliton solution of the coupled KP equation is given by τ^{G}, σ^{G} and $\widehat{\sigma}^{\mathrm{G}}$.

Remarks

(1) The Gramm-type pfaffian τ^{G} may be transformed into a solution of the BKP hierarchy:

$$\begin{aligned}
&[(D_3 - D_1^3)D_{-1} + 3D_1^2]\tau \cdot \tau = 0, \\
&[D_1^6 - 5D_1^3D_3 - 5D_3^2 + 9D_1D_5]\tau \cdot \tau = 0, \qquad (3.101) \\
&\qquad \ldots
\end{aligned}$$

In this case, choosing functions f_i, g_i to be

$$f_i = \frac{\partial}{\partial x}\phi_i, \quad g_i = \phi_i, \tag{3.102}$$

we have

$$(i, j) = c_{ij} + \int^x D_x\phi_i \cdot \phi_j \,\mathrm{d}x, \tag{3.103}$$

which coincides with the (i, j) entry of the pfaffian solution of the BKP hierarchy.

(2) In order to obtain a solution of the KP hierarchy, we choose functions f_i satisfying (3.100) and

$$g_i = 0\,(1 \le i \le M), \quad c_{ij} = 0\,(1 \le i < j \le M) \tag{3.104}$$

for an arbitrary natural number M. Then we have

$$(i, j) = \begin{cases} 0 & (1 \le i < j \le M) \\ c_{ij} + \int^x f_i g_j \,\mathrm{d}x & (1 \le i \le M,\ M+1 \le j \le 2N). \end{cases} \tag{3.105}$$

If (i) $M = N$, $\tau^{\mathrm{G}} = (1, 2, \ldots, N, N+1, \ldots, 2N)$ gives an $N \times N$ grammian,

$$\tau^{\mathrm{G}} = \det(m_{ij})_{1\le i<j\le N}, \tag{3.106}$$
$$m_{ij} = c_{i,2N+1-j} + \int^x f_j g_{2N+1-j} \,\mathrm{d}x,$$

which gives the grammian solution of the KP equation. If (ii) $M > N$, we have $\tau^{\mathrm{G}} = 0$, and otherwise (iii) $M < N$, we have a hybrid mode solution in which the solution of the KP equation and that of the coupled KP equation coexist. □

3.5 The two-dimensional Toda lattice equation

3.5.1 Wronskian solutions

The *two-dimensional Toda lattice equation* is the system

$$\frac{\partial^2 Q_n}{\partial s \partial x} = V_{n+1} - 2V_n + V_{n-1}, \tag{3.107}$$

$$Q_n = \log(1 + V_n), \tag{3.108}$$

where $n = \ldots, -1, 0, 1, \ldots$. Through the dependent variable transformation

$$V_n = \frac{\partial^2}{\partial s \partial x} \log(\tau_n), \tag{3.109}$$

(3.107) may be integrated with respect to x and s to obtain

$$1 + \frac{\partial^2}{\partial s \partial x} \log(\tau_n) = \frac{\tau_{n+1}\tau_{n-1}}{\tau_n^2}, \tag{3.110}$$

where we have set constants of integration equal to zero. Expanding the derivative of $\log(\tau_n)$ and clearing fractions, we obtain the bilinear equation

$$\frac{\partial^2 \tau_n}{\partial s \partial x}\tau_n - \frac{\partial \tau_n}{\partial s}\frac{\partial \tau_n}{\partial x} = \tau_{n+1}\tau_{n-1} - \tau_n^2, \tag{3.111}$$

which is written in terms of D-operators as

$$D_x D_s \tau_n \cdot \tau_n = 2(\tau_{n+1}\tau_{n-1} - \tau_n^2). \tag{3.112}$$

Remark

The one-dimensional Toda lattice equation is given by

$$\frac{\partial^2 Q_n}{\partial t^2} = V_{n+1} - 2V_n + V_{n-1}, \tag{3.113}$$

$$Q_n = \log(1 + V_n). \tag{3.114}$$

The system that is nowadays referred to as the two-dimensional Toda lattice is obtained by considering a two-dimensional version of the term $\partial^2 Q_n/\partial t^2$ on the left-hand side of (3.113). In physical terms, a two-dimensional Toda equation should be obtained by considering a two-dimensional version of the force term on the right-hand side. However, up to now we have found an N-soliton only for the system (3.107), (3.108). This is why we call this system the two-dimensional Toda lattice. □

The N-soliton solution for (3.112) is expressed as a wronskian [49]

$$\tau_n = \begin{vmatrix} \phi_1(n) & \phi_1(n+1) & \cdots & \phi_1(n+N-1) \\ \phi_2(n) & \phi_2(n+1) & \cdots & \phi_2(n+N-1) \\ \vdots & \vdots & & \vdots \\ \phi_N(n) & \phi_N(n+1) & \cdots & \phi_N(n+N-1) \end{vmatrix}, \tag{3.115}$$

where each $\phi_i(n)$ satisfies the differential equations

$$\frac{\partial \phi_i(n)}{\partial x} = \phi_i(n+1), \qquad \frac{\partial \phi_i(n)}{\partial s} = -\phi_i(n-1). \tag{3.116}$$

Remark

The above determinant is simply the $N \times N$ Casorati determinant appearing in the theory of linear difference equations. □

Let us now prove that τ_n given by (3.116) satisfies (3.112). As in the KP equation case, we express τ_n in terms of a Maya diagram,

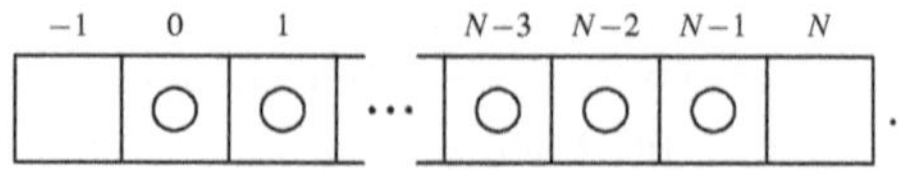

Its derivatives may also be expressed in terms of Maya diagrams as

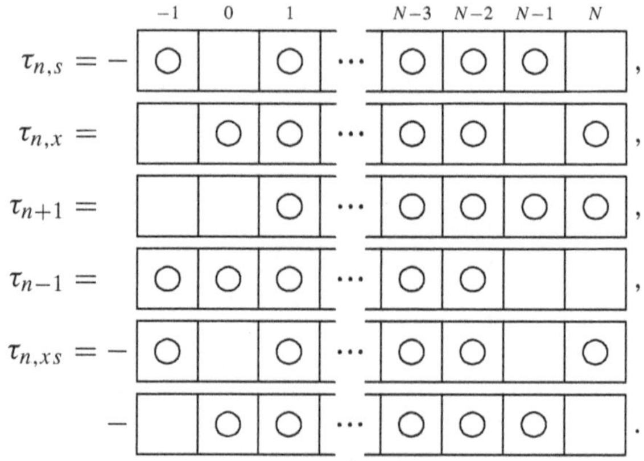

Substituting the above results into the bilinear equation

$$\frac{\partial^2 \tau_n}{\partial s \partial x}\tau_n - \frac{\partial \tau_n}{\partial s}\frac{\partial \tau_n}{\partial x} = \tau_{n+1}\tau_{n-1} - \tau_n^2 \tag{3.117}$$

gives the Maya diagram expression

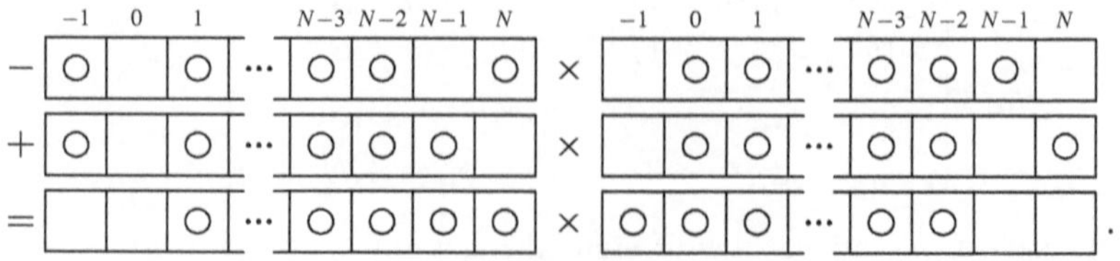

Omitting boxes $1, 2, \ldots, N-2$, which contain particles in all of the above terms, gives

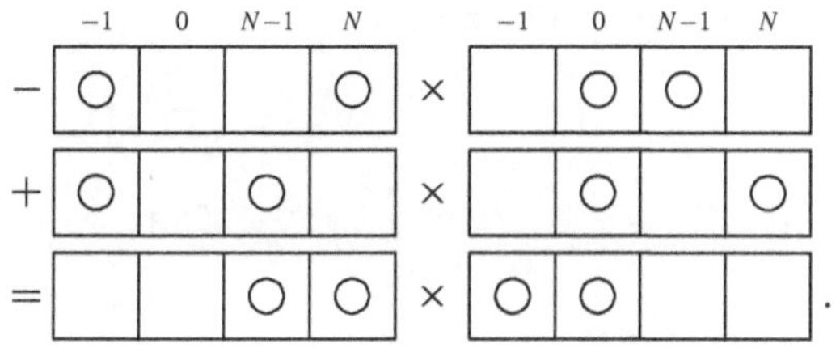

This is simply the Plücker relation for determinants, and therefore is automatically satisfied.

3.5.2 Grammian solutions

The two-dimensional Toda lattice equation in bilinear form,

$$\frac{\partial^2 \tau_n}{\partial s \partial x}\tau_n - \frac{\partial \tau_n}{\partial s}\frac{\partial \tau_n}{\partial x} = \tau_{n+1}\tau_{n-1} - \tau_n^2, \tag{3.118}$$

has a solution τ_n expressed as a grammian [50]:

$$\tau_n = \left| c_{ij} + (-)^n \int^x f_i^{(n)} g_j^{(-n)} \, \mathrm{d}x \right|_{1 \le i,j \le N}. \tag{3.119}$$

In the above equation, each $f_i^{(n)}$, $g_j^{(-n)}$ satisfy the linear differential equations

$$\begin{aligned} &\frac{\partial f_i^{(n)}}{\partial x_k} = f_i^{(n+k)}, \quad \frac{\partial g_i^{(-n)}}{\partial x_k} = (-)^{k-1} g_i^{(-n+k)}, \\ &\frac{\partial f_i^{(n)}}{\partial s_k} = -f_i^{(n-k)}, \quad \frac{\partial g_i^{(n)}}{\partial s_k} = (-)^k g_i^{(-n-k)}, \\ &x_1 = x, \; s_1 = s, \quad k = 1, 2, \ldots, \end{aligned} \tag{3.120}$$

where c_{ij} is constant and we write $(-1)^n$ as $(-)^n$ for short.

Remarks

(1) The superscript (n) in $f_i^{(n)}$ denotes the nth derivative (or $-n$th antiderivative if $n < 0$) with respect to x.

(2) In (3.119), τ_n is exactly the same as the grammian expression for the solution of the KP equation apart from the factor $(-)^n$. This factor is not important for solutions of the two-dimensional Toda lattice equation, but is necessary to discuss the connection with the BKP solution. □

Let us express τ_n by means of a pfaffian:

$$\tau_n = (1, 2, \dots, N, N^*, \dots, 2^*, 1^*)_n,$$

$$(i, j^*)_n = c_{ij} + (-)^n \int^x f_i^{(n)} g_j^{(-n)} \, \mathrm{d}x,$$

$$(i, j)_n = (i^*, j^*)_n = 0.$$

The derivatives with respect to x and s and shifts in n of the pfaffian entry $(i, j^*)_n$ are as follows:

$$\begin{aligned}
\frac{\partial}{\partial x}(i, j^*)_n &= (-)^n f_i^{(n)} g_j^{(-n)} \\
&= (d_{-n}, d_n^*, i, j^*)_n, \\
\frac{\partial}{\partial s}(i, j^*)_n &= (-)^{n-1} f_i^{(n-1)} g_j^{(-n-1)} \\
&= (d_{-n-1}, d_{n-1}^*, i, j^*)_n, \\
\frac{\partial^2}{\partial x \partial s}(i, j^*)_n &= (-)^{n-1} [f_i^{(n)} g_j^{(-n-1)} + f_i^{(n-1)} g_j^{(-n)}] \\
&= (d_{-n-1}, d_n^*, i, j^*)_n - (d_{-n}, d_{n-1}^*, i, j^*)_n, \\
(i, j^*)_{n+1} &= c_{ij} + (-)^{n+1} \int^x f_i^{(n+1)} g_j^{(-n-1)} \, \mathrm{d}x \\
&= c_{ij} + (-)^n \int^x f_i^{(n)} g_j^{(-n)} \, \mathrm{d}x + (-)^{n-1} f_i^{(n)} g_j^{(-n-1)} \\
&= (i, j^*)_n + (d_{-n-1}, d_n^*, i, j^*)_n, \\
(i, j^*)_{n-1} &= c_{ij} + (-)^{n-1} \int^x f_i^{(n-1)} g_j^{(-n+1)} \, \mathrm{d}x \\
&= c_{ij} + (-)^n \int^x f_i^{(n)} g_j^{(-n)} \, \mathrm{d}x + (-)^{n-1} f_i^{(n-1)} g_j^{(-n)} \\
&= (i, j^*)_n - (d_{-n}, d_{n-1}^*, i, j^*)_n,
\end{aligned}$$

where

$$(d_n^*, i)_n = f_i^{(n)},$$

$$(d_{-n}, j^*) = (-)^n g_j^{(-n)},$$

$$(d_{-n}, d_n^*) = (d_{-n}, d_n) = (d_{-n}^*, d_n^*) = 0.$$

As a result, rewriting $\tau_n = (\ldots)_n$, we obtain

$$\frac{\partial}{\partial x}(\ldots)_n = (d_{-n}, d_n^*, \ldots)_n,$$

$$\frac{\partial}{\partial s}(\ldots)_n = (d_{-n-1}, d_{n-1}^*, \ldots)_n,$$

$$\frac{\partial^2}{\partial x \partial s}(\ldots)_n = (d_{-n-1}, d_n^*, \ldots)_n - (d_{-n}, d_{n-1}^*, \ldots)_n$$
$$+ (d_{-n-1}, d_{n-1}^*, d_{-n}, d_n^*, \ldots)_n,$$

$$\tau_{n+1} = (\ldots)_n + (d_{-n-1}, d_n^*, \ldots)_n,$$

$$\tau_{n-1} = (\ldots)_n - (d_{-n}, d_{n-1}^*, \ldots)_n.$$

By employing the above equations, the two-dimensional Toda lattice equation in bilinear form,

$$\frac{\partial^2 \tau_n}{\partial s \partial x}\tau_n - \frac{\partial \tau_n}{\partial s}\frac{\partial \tau_n}{\partial x} = \tau_{n+1}\tau_{n-1} - \tau_n^2, \tag{3.121}$$

is expressed by means of Maya diagrams as

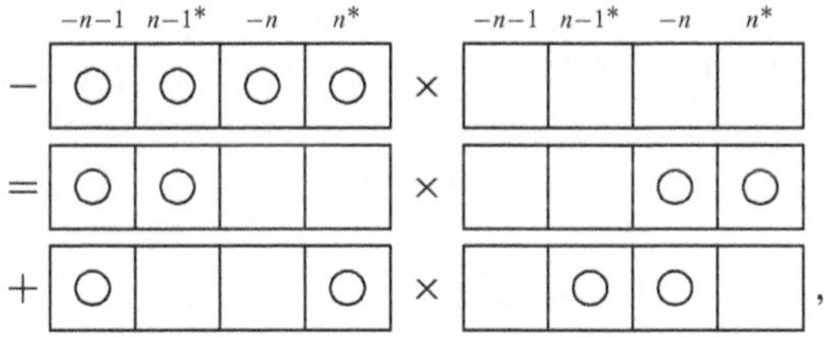

which shows that τ_n satisfies the bilinear equation (3.121).

Next, we discuss the BKP version of the grammian τ_n. We note that

$$\tau_n = \det((m_{ij})_n), \tag{3.122}$$
$$(m_{ij})_n = c_{ij} + \int^x f_i^{(n)} g_j^{(-n)}\, \mathrm{d}x$$

solves the KP equation if $n = 0$. We here assume that f_i, g_j satisfy

$$g_i = -2\frac{\partial}{\partial x} f_i, \tag{3.123}$$

where $x = x_1$.

Then, if $c_{ij} = -c_{ji}$, each entry in τ_n,

$$(m_{ij})_n \equiv c_{ij} - (-)^n 2 \int^x f_i^{(n)} f_j^{(1-n)}\, \mathrm{d}x, \tag{3.124}$$

satisfies the relation

$$(m_{ij})_{1-n} = -(m_{ij})_n. \tag{3.125}$$

Hence, we have

$$\tau_{1-n} = (-)^N \tau_n, \tag{3.126}$$

where N is the size of the determinant. In the terminology of Jimbo and Miwa [15], this is the BKP version of τ_n. This is because $(m_{ij})_0$ can be written as

$$(m_{ij})_0 = c_{ij} + \int^x D_x f_i \cdot f_j\, \mathrm{d}x - f_i f_j \tag{3.127}$$

after integration by parts. Employing the relation (2.92), that is

$$\begin{vmatrix} a_{11} - y_1 y_1 & a_{12} - y_1 y_2 & \cdots & a_{1n} - y_1 y_n \\ a_{21} - y_2 y_1 & a_{22} - y_2 y_2 & \cdots & a_{2n} - y_2 y_n \\ a_{31} - y_3 y_1 & a_{32} - y_3 y_2 & \cdots & a_{3n} - y_3 y_n \\ \vdots & \vdots & & \vdots \\ a_{n1} - y_n y_1 & a_{12} - y_n y_2 & \cdots & a_{nn} - y_n y_n \end{vmatrix} = (1, 2, \cdots, n)^2,$$

where $(i, j) \equiv a_{ij}$ and n is even, we have

$$\tau_0 = \tau_{\mathrm{BKP}}^2. \tag{3.128}$$

Remark

The linear differential equations (3.120), satisfied by f_j and g_j, are compatible with the relation

$$g_j = -2\frac{\partial}{\partial x} f_j \quad (x = x_1) \tag{3.129}$$

only when k is odd. Therefore, we must freeze the dependence on x_k and s_k when k is even. There is no problem in doing this because the BKP equation is written only in terms of odd index variables. □

3.6 The two-dimensional Toda molecule equation

3.6.1 Bi-directional wronskian solutions

The two-dimensional Toda molecule equation is written as

$$\frac{\partial^2 Q_n}{\partial x \partial y} = V_{n+1} - 2V_n + V_{n-1}, \tag{3.130}$$

$$Q_n \equiv \log(V_n). \tag{3.131}$$

We note that it has the same form as the Toda lattice equation except that for the Toda lattice equation

$$Q_n \equiv \log(1 + V_n). \tag{3.132}$$

Therefore, if $V_n = 0$ the lattice equation has $Q_n = 0$, whereas the molecule equation has $Q_n = -\infty$. In the Toda molecule equation (3.130), (3.131), the independent variable n takes values $n = 1, 2, \ldots, N$ and its boundary condition is given by

$$V_0 = V_{N+1} = 0. \tag{3.133}$$

This difference in the boundary conditions affects the structure of its solutions.

Through the same dependent variable transformation,

$$V_n = \frac{\partial^2}{\partial x \partial y} \log(\tau_n), \tag{3.134}$$

(3.130) can be integrated with respect to x and y to give

$$\frac{\partial^2}{\partial x \partial y} \log(\tau_n) = \frac{\tau_{n+1}\tau_{n-1}}{\tau_n^2}, \tag{3.135}$$

where the constant of integration has been set equal to zero. Expanding the derivatives and clearing fractions, we obtain the bilinear form,

$$\frac{\partial^2 \tau_n}{\partial x \partial y}\tau_n - \frac{\partial \tau_n}{\partial x}\frac{\partial \tau_n}{\partial y} = \tau_{n+1}\tau_{n-1}, \tag{3.136}$$

or, equivalently,

$$D_x D_y \tau_n \cdot \tau_n = 2\tau_{n+1}\tau_{n-1}. \tag{3.137}$$

Employing (3.135), we have

$$V_n = \frac{\partial^2}{\partial x \partial y} \log(\tau_n) = \frac{\tau_{n+1}\tau_{n-1}}{\tau_n^2}. \tag{3.138}$$

By virtue of (3.134), the boundary conditions $V_0 = V_{N+1} = 0$ are satisfied by choosing

$$\tau_0 = 1, \ \tau_{N+1} = \Phi(x)\chi(y), \tag{3.139}$$

where $\Phi(x)$ and $\chi(y)$ are arbitrary functions in x and y, respectively. We use the convention

$$\tau_{-1} = 0, \ \tau_{N+2} = 0 \tag{3.140}$$

so that (3.138) is satisfied for $n = 0$ and $n = N + 1$.

The solution τ_n of the bilinear equation is expressed by means of an $n \times n$ wronskian [51],

$$\tau_0 = 1, \quad \tau_n = \left| \left(\frac{\partial}{\partial x}\right)^{i-1} \left(\frac{\partial}{\partial y}\right)^{j-1} \Psi(x, y) \right|_{1 \le i, j \le n}, \tag{3.141}$$

where $\Psi(x, y)$ is, for now, an arbitrary function of x, y and the natural number n is not only the position in the Toda molecule, but also the degree of the wronskian. The above determinant is a wronskian in both the rows and the columns and is called a *bi-directional wronskian*. Explicit forms of the first few of the τ_n are given by

$$\tau_1 = \Psi_{00}, \quad \tau_2 = \begin{vmatrix} \Psi_{00} & \Psi_{01} \\ \Psi_{10} & \Psi_{11} \end{vmatrix}, \quad \tau_3 = \begin{vmatrix} \Psi_{00} & \Psi_{01} & \Psi_{02} \\ \Psi_{10} & \Psi_{11} & \Psi_{12} \\ \Psi_{20} & \Psi_{21} & \Psi_{22} \end{vmatrix}, \tag{3.142}$$

where we have adopted the notation

$$\Psi_{ij} = \left(\frac{\partial}{\partial x}\right)^i \left(\frac{\partial}{\partial y}\right)^j \Psi(x, y). \tag{3.143}$$

Let us confirm that the bilinear equation (3.136) holds in the cases $n = 1$ and $n = 2$. If $n = 1$, the bilinear equation

$$\tau_{1,xy}\tau_1 - \tau_{1,x}\tau_{1,y} = \tau_2\tau_0 \tag{3.144}$$

is

$$\Psi_{11}\Psi_{00} - \Psi_{10}\Psi_{01} = \begin{vmatrix} \Psi_{00} & \Psi_{01} \\ \Psi_{10} & \Psi_{11} \end{vmatrix}. \tag{3.145}$$

If $n = 2$, the bilinear equation

$$\tau_{2,xy}\tau_2 - \tau_{2,x}\tau_{2,y} = \tau_3\tau_1 \tag{3.146}$$

is

$$\begin{vmatrix} \Psi_{00} & \Psi_{02} \\ \Psi_{20} & \Psi_{22} \end{vmatrix}\begin{vmatrix} \Psi_{00} & \Psi_{01} \\ \Psi_{10} & \Psi_{11} \end{vmatrix} - \begin{vmatrix} \Psi_{00} & \Psi_{01} \\ \Psi_{20} & \Psi_{21} \end{vmatrix}\begin{vmatrix} \Psi_{00} & \Psi_{02} \\ \Psi_{10} & \Psi_{12} \end{vmatrix}$$
$$= \begin{vmatrix} \Psi_{00} & \Psi_{01} & \Psi_{02} \\ \Psi_{10} & \Psi_{11} & \Psi_{12} \\ \Psi_{20} & \Psi_{21} & \Psi_{22} \end{vmatrix} \Psi_{00}. \tag{3.147}$$

In order to prove that τ_n given by (3.141) solves the bilinear equation (3.136), we introduce $(n+1) \times (n+1)$, $n \times n$ and $(n-1) \times (n-1)$ determinants D, $D\begin{bmatrix} i \\ j \end{bmatrix}$ and $D\begin{bmatrix} i & j \\ k & l \end{bmatrix}$:

$$D \equiv \left| \left(\frac{\partial}{\partial x}\right)^{i-1} \left(\frac{\partial}{\partial y}\right)^{j-1} \Psi(x, y) \right|_{1\le i,j\le n+1} = \tau_{n+1}, \tag{3.148a}$$

$$D\begin{bmatrix} i \\ j \end{bmatrix} = \text{determinant obtained by eliminating the } i\text{th row and } j\text{th column of } D, \tag{3.148b}$$

$$D\begin{bmatrix} i & j \\ k & l \end{bmatrix} = \text{determinant obtained by eliminating the } i\text{th and } j\text{th rows and the } k\text{th and } l\text{th columns of } D. \tag{3.148c}$$

Using the above notation, we have

$$\tau_n = D\begin{bmatrix} n+1 \\ n+1 \end{bmatrix}, \tag{3.149a}$$

$$\tau_{n-1} = D\begin{bmatrix} n & n+1 \\ n & n+1 \end{bmatrix}, \tag{3.149b}$$

$$\frac{\partial \tau_n}{\partial x} = D\begin{bmatrix} n \\ n+1 \end{bmatrix}, \tag{3.149c}$$

$$\frac{\partial \tau_n}{\partial y} = D\begin{bmatrix} n+1 \\ n \end{bmatrix}, \tag{3.149d}$$

$$\frac{\partial^2 \tau_n}{\partial x \partial y} = D\begin{bmatrix} n \\ n \end{bmatrix}, \tag{3.149e}$$

from which we see that the bilinear equation (3.136) is equivalent to

$$D\begin{bmatrix} n \\ n \end{bmatrix} D\begin{bmatrix} n+1 \\ n+1 \end{bmatrix} - D\begin{bmatrix} n \\ n+1 \end{bmatrix} D\begin{bmatrix} n+1 \\ n \end{bmatrix} = D\begin{bmatrix} n & n+1 \\ n & n+1 \end{bmatrix} D. \tag{3.150}$$

This is simply the Jacobi identity for determinants. Hence, we have verified that τ_n is a solution of (3.136).

The arbitrary function $\Psi(x, y)$ needs to satisfy the boundary condition $\tau_{N+1} = \Phi(x)\chi(y)$. To this end, we introduce arbitrary functions $u_j(x)$ and $v_j(y)$ $(j = 1, 2, \ldots, N+1)$, of x and y, respectively, and put

$$\Psi(x, y) = \sum_{j=1}^{N+1} u_j(x) v_j(y). \tag{3.151}$$

Choosing $\Psi(x, y)$ as above, we have

$$\begin{aligned} \tau_n &= \left| \left(\frac{\partial}{\partial x}\right)^{i-1} \left(\frac{\partial}{\partial y}\right)^{j-1} \Psi(x, y) \right|_{1 \le i, j \le n} \\ &= \left| \sum_{j=1}^{N+1} \left(\frac{\partial}{\partial x}\right)^{i-1} u_j(x) \left(\frac{\partial}{\partial y}\right)^{k-1} v_j(y) \right|_{1 \le i, k \le n}. \end{aligned} \tag{3.152}$$

This is equal to the determinant of the product of the $n \times (N+1)$ matrix A_n and the $(N+1) \times n$ matrix B_n,

$$\tau_n = |A_n \times B_n|, \tag{3.153}$$

where

$$(A_n)_{ij} = \frac{\partial^{i-1}}{\partial x^{i-1}} u_j(x) \quad (B_n)_{jk} = \frac{\partial^{k-1}}{\partial y^{k-1}} v_j(y). \tag{3.154}$$

In the case $n = N + 1$, A_n and B_n are square matrices and therefore

$$\tau_{N+1} = |A_{N+1}| \times |B_{N+1}|. \tag{3.155}$$

Since $|A_{N+1}|$ and $|B_{N+1}|$ depend only on x and y, respectively, τ_{N+1} may be rewritten as

$$\tau_{N+1} = \Phi(x)\chi(y), \tag{3.156}$$

which is a boundary condition for τ_n. This verifies that τ_n is a solution for the Toda molecule equation satisfying the boundary condition.

Remarks

(1) Solutions of the Toda molecule equation were first found by Leznov and Savaliev [52], using a group theoretical approach. This is, however, far beyond a beginner's understanding and so we have employed a method using wronskians.

(2) We note that the two-dimensional Toda molecule equation,

$$\frac{\partial^2 Q_n}{\partial x \partial y} = V_{n+1} - 2V_n + V_{n-1}, \tag{3.157}$$

$$Q_n \equiv \log(V_n), \tag{3.158}$$

includes the Liouville equation discussed in Chapter 1 as a special case. By choosing the boundary condition $V_0 = V_2 = 0$, the above equation is equivalent to

$$\begin{aligned} \frac{\partial^2 Q_1}{\partial x \partial y} &= -2V_1 \\ &= \exp(Q_1), \end{aligned} \tag{3.159}$$

which is simply the Liouville equation.

(3) In Chapter 1 we discussed the two-wave interaction equation,

$$\frac{\partial \phi_1}{\partial \xi} = -\phi_1\phi_2, \quad \frac{\partial \phi_2}{\partial \eta} = \phi_1\phi_2, \tag{3.160}$$

which has a form similar to the Liouville equation. In fact, this equation is generated from the Bäcklund transformation of the Toda molecule equation [53]. □

3.6.2 Double wronskian solutions

In this section, we consider another expression for the solution of the bilinear two-dimensional Toda molecule equation,

$$D_x D_y \tau_n \cdot \tau_n = 2\tau_{n+1}\tau_{n-1}, \tag{3.161}$$

where we set the boundary conditions $\tau_0 = \chi(y)$, $\tau_{N+1} = \Phi(x)$. To this end, we introduce arbitrary functions $f_i(x)$ and $g_i(y)$ $(i = 1, 2, \dots, M)$, which depend only on x and y, respectively, and consider the $M \times M$ *double wronskian* [54]

$$\tau_n = \begin{vmatrix} f_1^{(0)} & f_1^{(1)} & \cdots & f_1^{(n-1)} & g_1^{(0)} & g_1^{(1)} & \cdots & g_1^{(M-n-1)} \\ f_2^{(0)} & f_2^{(1)} & \cdots & f_2^{(n-1)} & g_2^{(0)} & g_2^{(1)} & \cdots & g_2^{(M-n-1)} \\ \vdots & \vdots & & \vdots & \vdots & \vdots & & \vdots \\ f_M^{(0)} & f_M^{(1)} & \cdots & f_M^{(n-1)} & g_M^{(0)} & g_M^{(1)} & \cdots & g_M^{(M-n-1)} \end{vmatrix}, \tag{3.162}$$

where

$$\begin{aligned} f_i^{(n)} &= \frac{\partial^n}{\partial x^n} f_i, & g_i^{(n)} &= \frac{\partial^n}{\partial y^n} g_i, \\ \frac{\partial}{\partial x_n} f_i &= \frac{\partial^n}{\partial x^n} f_i, & \frac{\partial}{\partial y_n} g_i &= \frac{\partial^n}{\partial y^n} g_i, \end{aligned} \tag{3.163}$$

for $i = 1, 2, \dots, M,\ n = 1, 2, \dots$. From this, we have

$$\tau_0 = \begin{vmatrix} g_1^{(0)} & g_1^{(1)} & \cdots & g_1^{(M-1)} \\ g_2^{(0)} & g_2^{(1)} & \cdots & g_2^{(M-1)} \\ \vdots & \vdots & & \vdots \\ g_M^{(0)} & g_M^{(1)} & \cdots & g_M^{(M-1)} \end{vmatrix} = \chi(y) \tag{3.164}$$

and

$$\tau_M = \begin{vmatrix} f_1^{(0)} & f_1^{(1)} & \cdots & f_1^{(M-1)} \\ f_2^{(0)} & f_2^{(1)} & \cdots & f_2^{(M-1)} \\ \vdots & \vdots & & \vdots \\ f_M^{(0)} & f_M^{(1)} & \cdots & f_M^{(M-1)} \end{vmatrix} = \Phi(y), \tag{3.165}$$

from which we see that the boundary conditions are automatically satisfied.

Remarks

(1) τ_n is also a solution for the KP hierarchy with independent variables $x_1, x_2, x_3, \dots$ or $y_1, y_2, y_3, \dots$.

(2) The two-dimensional Toda molecule equation with $M = \infty$ appears in a book by G. Darboux [55]. The τ-function expressed in double wronskian form can also be observed in this book. This was first noticed by Yoshinori Kametaka [56]. In this paper, he discussed hypergeometric solutions of the Toda molecule equation.

(3) Generalization of a double wronskian to a triple wronskian, which is made from arbitrary functions in independent variables x_1, y_1, z_1 given by $f(x_1), g(y_1), h(z_1)$, yields the six-wave interaction equation (a generalization of the three-wave interaction equation). □

If we express τ_n by means of a Maya diagram,

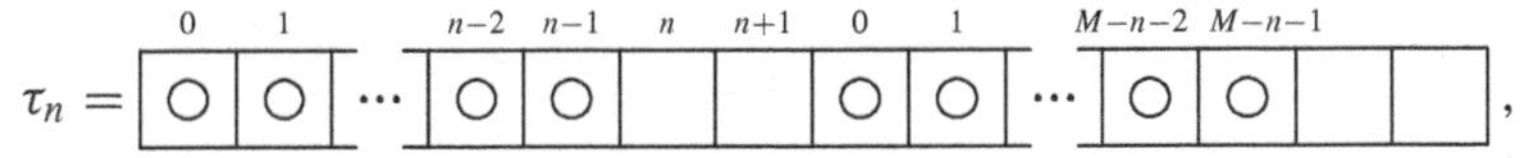

the derivatives of τ_n and $\tau_{n\pm1}$ are expressed as

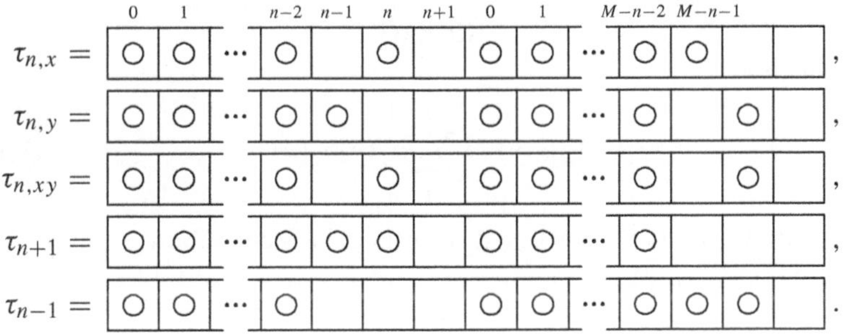

By re-ordering the boxes in the above Maya diagrams, we obtain

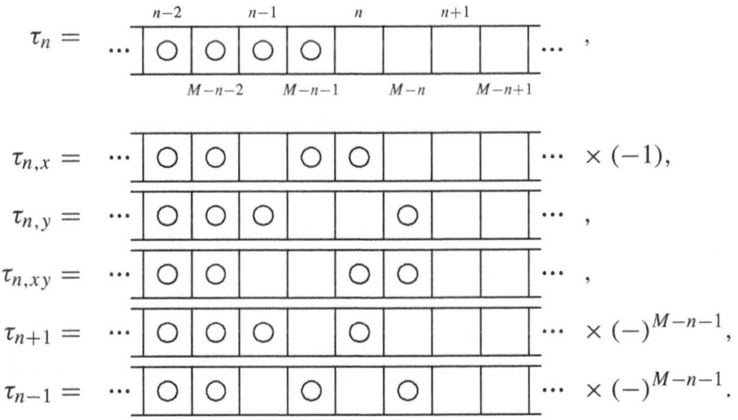

By using these Maya diagrams, the bilinear two-dimensional Toda molecule equation,

$$\tau_n \tau_{n,xy} - \tau_{n+1}\tau_{n-1} - \tau_{n,y}\tau_{n,x} = 0, \tag{3.166}$$

is rewritten as

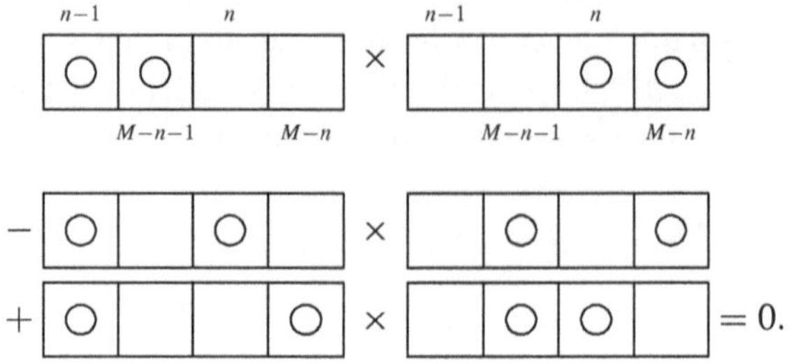

This shows that τ_n satisfies the two-dimensional Toda molecule equation.

Remark

The function

$$\Psi(x, y) = \sum_{j=1}^{M} u_j(x) v_j(y), \tag{3.167}$$

which appears in the bi-directional wronskian in Section 3.6.1, is given by the ratio of τ_1 and τ_0 [54]:

$$\Psi(x, y) = \frac{\tau_1}{\tau_0}. \tag{3.168}$$

□

4
Bäcklund transformations

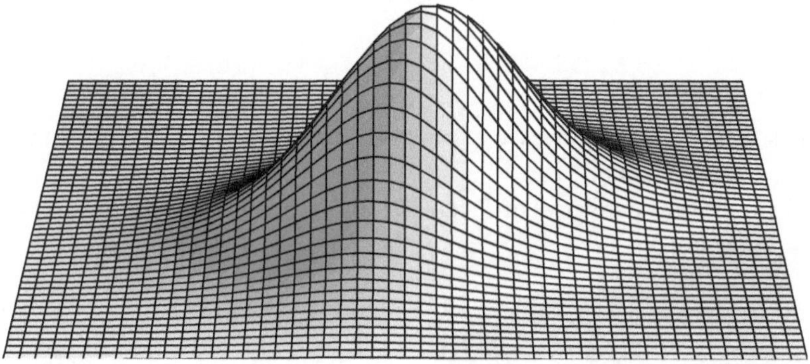

Dromion.

4.1 What is a Bäcklund transformation?

A Bäcklund transformation is a transformation between a solution u of a given linear or nonlinear differential equation,

$$L_1(u, u_t, u_x, u_{xx}, u_{xxx}, u_y, \ldots) = 0,$$

and another solution v of another differential equation,

$$L_2(v, v_t, v_x, v_{xx}, v_{xxx}, v_y, \ldots) = 0,$$

which may be the same as, or different from, L_1.

As a simple example, we consider the dispersionless KdV and mKdV equations,

$$u_t + uu_x = 0, \tag{4.1}$$

$$v_t + (av^2 + bv)v_x = 0, \tag{4.2}$$

where a, b are constants. These equations coincide if $b = 1$ and $a = 0$. A Bäcklund transformation between solutions of (4.1) and (4.2) is given by

$$u = av^2 + bv. \tag{4.3}$$

This is confirmed by taking the t-derivative of (4.3) and using (4.2) to obtain

$$u_t = (av^2 + bv)_t = -(2av + b)(av^2 + bv)v_x = -uu_x.$$

Since (4.2) may be written as

$$v_t + \left(\frac{1}{3}av^3 + \frac{1}{2}bv^2\right)_x = 0, \tag{4.4}$$

v is a conserved density, or, in other words, $\int v\,\mathrm{d}x$ is a conserved quantity.

Remarks

(1) In (4.3), u is expressed explicitly in terms of v and is called the Miura transformation.

(2) If the time derivative of T can be expressed as the space derivative of some quantity X,

$$T_t + X_x = 0,$$

then T is called a *conserved density*. Taking the t-derivative of the integral of T over an interval $[a, b]$, we have

$$\left\{\int_a^b T\,\mathrm{d}x\right\}_t = -\int_a^b X_x\,\mathrm{d}x = X(a) - X(b).$$

If we impose the boundary condition $X(a) = X(b)$, the integral of T over $[a, b]$ is independent of t, and is called a *conserved quantity*. □

Next, from (4.3), v may be rewritten in terms of u as

$$v = \frac{1}{2a}(-b \pm (b^2 + 4au)^{1/2}).$$

Taking $b = 1$ and the upper sign, the above equation may be expanded for sufficiently small a as

$$v \sim u - au^2 + 2a^2u^3 - \cdots.$$

Since v is a conserved density, that is it satisfies (4.4), each coefficient in the expansion of v with respect to a is again a conserved quantity. This means that every power of u is a conserved quantity. We can of course check this fact directly using (4.2). However, it is interesting to note that an infinite number of conserved quantities are generated from a Bäcklund transformation containing an arbitrary parameter [57].

We showed in Chapter 1 that the Liouville equation,

$$u_{xy} = \mathrm{e}^{u}, \tag{4.5}$$

may be reduced to the linear equation,

$$v_{xy} = 0. \tag{4.6}$$

Differentiating the equations

$$u_x + v_x = 2^{1/2}\mathrm{e}^{(u-v)/2}, \tag{4.7a}$$

$$u_y - v_y = 2^{1/2}\mathrm{e}^{(u+v)/2} \tag{4.7b}$$

with respect to y and x, respectively, we obtain

$$u_{xy} + v_{xy} = 2^{-1/2}(u_y - v_y)\mathrm{e}^{(u-v)/2} = \mathrm{e}^{u},$$

$$u_{yx} - v_{yx} = 2^{-1/2}(u_x + v_x)\mathrm{e}^{(u+v)/2} = \mathrm{e}^{u}.$$

Adding and subtracting these equations gives the Liouville equation (4.5) and the linear equation (4.6), respectively. This shows that (4.7) is a Bäcklund transformation connecting the solutions of these two equations [58].

The first Bäcklund transformation to be found in connection with soliton theory is related to the sine–Gordon equation,

$$u_{xy} = \sin u. \tag{4.8}$$

This Bäcklund transformation is given by [58]

$$\frac{1}{2}(u+v)_x = p \sin\left[\frac{1}{2}(u-v)\right], \tag{4.9a}$$

$$\frac{1}{2}(u-v)_y = p^{-1} \sin\left[\frac{1}{2}(u+v)\right]. \tag{4.9b}$$

Differentiating the above equations with respect to y and x, respectively, we have

$$\frac{1}{2}(u+v)_{xy} = \frac{1}{2}p(u-y)_y \cos\left[\frac{1}{2}(u-v)\right]$$
$$= \sin\left[\frac{1}{2}(u+v)\right]\cos\left[\frac{1}{2}(u-v)\right],$$
$$\frac{1}{2}(u-v)_{yx} = \frac{1}{2}p^{-1}(u+y)_x \cos\left[\frac{1}{2}(u+v)\right]$$
$$= \sin\left[\frac{1}{2}(u-v)\right]\cos\left[\frac{1}{2}(u+v)\right].$$

Addition and subtraction of these both give a copy of the sine–Gordon equation,

$$u_{xy} = \sin u,$$
$$v_{xy} = \sin v.$$

Therefore, (4.9) is a Bäcklund transformation between two solutions of the sine–Gordon equation. The Bäcklund transformation may be used to obtain the N-soliton solution and an infinite number of conserved quantities [57] of soliton equations. A relation between this Bäcklund transformation and the inverse scattering method is also known [59, 60].

As was shown in Chapter 1, the sine–Gordon equation,

$$u_{xy} = \sin u,$$

reduces to the bilinear form

$$D_x D_y f \cdot f - \frac{1}{2}(f^2 - f^{*2}) = 0,$$

through the bi-logarithmic transformation

$$u = 2\mathrm{i}\log(f/f^*).$$

It is interesting to determine the bilinear forms of (4.9a) and (4.9b). Here we simply present the result. The required bilinear forms are

$$D_x f \cdot g = -\frac{1}{2}p f^* g^*, \tag{4.10a}$$

$$D_y f \cdot g^* = -\frac{1}{2}p^{-1} f^* g, \tag{4.10b}$$

where g is another solution for the sine–Gordon equation, satisfying

$$D_x D_y g \cdot g - \frac{1}{2}(g^2 - g^{*2}) = 0. \tag{4.11}$$

The rest of this chapter is concerned with Bäcklund transformations in bilinear form [58,59].

4.2 Bäcklund transformations for KdV-type bilinear equations

First of all, we consider the Bäcklund transformation between a solution f for the general KdV-type bilinear equation

$$F(D_t, D_x, D_y) f \cdot f = 0, \tag{4.12}$$

and another solution f' for the same bilinear equation

$$F(D_t, D_x, D_y) f' \cdot f' = 0. \tag{4.13}$$

We will consider

$$P \equiv [F(D_t, D_x, D_y) f' \cdot f'] f^2 - f'^2 [F(D_t, D_x, D_y) f \cdot f]. \tag{4.14}$$

If $P = 0$, then f solves the bilinear equation (4.12) if and only if f' solves the bilinear equation (4.13). If we can obtain from $P = 0$ a pair of bilinear equations

$$\begin{aligned} F_1(D_t, D_x, D_y) f' \cdot f = 0, \\ F_2(D_t, D_x, D_y) f' \cdot f = 0, \end{aligned} \tag{4.15}$$

in which f, f' have interchanged their positions compared with (4.12) and (4.13), then they provide the Bäcklund transformations we are seeking.

We can use the *exchange formula*

$$\begin{aligned} &\exp(D_1)\left[\exp(D_2) a \cdot b\right] \cdot \left[\exp(D_3) c \cdot d\right] \\ &\quad = \exp\left(\frac{D_2 - D_3}{2}\right)\left[\exp\left(D_1 + \frac{D_2 + D_3}{2}\right) a \cdot d\right] \\ &\qquad \cdot \left[\exp\left(-D_1 + \frac{D_2 + D_3}{2}\right) c \cdot b\right], \end{aligned} \tag{4.16a}$$

where

$$D_i = \alpha_i D_t + \beta_i D_x + \gamma_i D_y \quad (i = 1, 2, 3), \tag{4.16b}$$

to carry out this interchange. As a typical example, let us find the Bäcklund transformation for the bilinear KdV equation [22, 59]

$$D_x(D_t + c_0 D_x + D_x^3) f \cdot f = 0, \tag{4.17}$$

where c_0 is a constant.

We start with

$$\begin{aligned} P \equiv & \left[D_x(D_t + c_0 D_x + D_x^3) f' \cdot f' \right] f^2 \\ & - f'^2 \left[D_x(D_t + c_0 D_x + D_x^3) f \cdot f \right]. \end{aligned} \tag{4.18}$$

Substituting

$$D_1 = \alpha D_x, \ D_2 = D_3 = \beta D_x, \ a = d, \ b = c \tag{4.19}$$

in the exchange formula (4.16), we obtain

$$\begin{aligned} & \left[\exp(\alpha D_x + \beta D_x) a \cdot a \right] \left[\exp(-\alpha D_x + \beta D_x) b \cdot b \right] \\ & = \exp(\alpha D_x) \left[\exp(\beta D_x) a \cdot b \right] \cdot \left[\exp(\beta D_x) b \cdot a \right]. \end{aligned} \tag{4.20}$$

Expanding the above equation in α, the coefficient of α^1 gives

$$\begin{aligned} & \left[\exp(\beta D_x) D_x a \cdot a \right] \left[\exp(\beta D_x) b \cdot b \right] - \left[\exp(\beta D_x) a \cdot a \right] \\ & \times \left[\exp(\beta D_x) D_x b \cdot b \right] = D_x \left[\exp(\beta D_x) a \cdot b \right] \cdot \left[\exp(\beta D_x) b \cdot a \right]. \end{aligned} \tag{4.21}$$

Taking a similar expansion in β, we obtain

$$\begin{aligned} \left[D_x^2 a \cdot a \right] b^2 - a^2 \left[D_x^2 b \cdot b \right] & = D_x \left[(D_x a \cdot b) \cdot ba + ab \cdot (D_x b \cdot a) \right] \\ & = 2 D_x (D_x a \cdot b) \cdot ba \end{aligned} \tag{4.22}$$

from the coefficient of β^1. Through the independent variable transformation $D_x \to D_x + \varepsilon D_t$, we have

$$\left[D_x D_t a \cdot a \right] b^2 - a^2 \left[D_x D_t b \cdot b \right] = 2 D_x (D_t a \cdot b) \cdot ba \tag{4.23}$$

from the coefficient of ε^1. Finally, the coefficient of β^3 in (4.21) gives

$$\begin{aligned}&\left[D_x^4 a\cdot a\right]b^2-a^2\left[D_x^4 b\cdot b\right]\\&\quad=2D_x\left[(D_x^3 a\cdot b)\cdot ba+3(D_x^2 a\cdot b)\cdot(D_x b\cdot a)\right].\end{aligned}\tag{4.24}$$

Substituting these equations into P, we have

$$\begin{aligned}P&=2D_x(D_t f'\cdot f)\cdot ff'+2c_0D_x(D_x f'\cdot f)\cdot ff'\\&\quad+2D_x\left[(D_x^3 f'\cdot f)\cdot ff'+3(D_x^2 f'\cdot f)\cdot(D_x f\cdot f')\right]\\&=2D_x\left[(D_t+c_0D_x+D_x^3)f'\cdot f\right]\cdot ff'\\&\quad+6D_x(D_x^2 f'\cdot f)\cdot(D_x f\cdot f').\end{aligned}\tag{4.25}$$

We may introduce two new arbitrary parameters λ and μ into the above equation to obtain

$$\begin{aligned}P=2D_x&\left\{\left[D_t+(c_0+3\lambda)D_x+D_x^3\right]f'\cdot f\right\}\cdot ff'\\&+6D_x\left[(D_x^2-\mu D_x-\lambda)f'\cdot f\right]\cdot(D_x f\cdot f').\end{aligned}\tag{4.26}$$

This is possible because the coefficients of λ and μ

$$\begin{aligned}\lambda:&\ D_x(D_x f'\cdot f)\cdot f'f+D_x f'f\cdot(D_x f'\cdot f),\\\mu:&\ -D_x(D_x f'\cdot f)\cdot(D_x f\cdot f'),\end{aligned}$$

are both equal to zero because of the property $D_x a\cdot b=-D_x b\cdot a$. Therefore, candidates for the Bäcklund transformation between f and f', which satisfy $P=0$, are

$$\left[D_t+(c_0+\lambda)D_x+D_x^3\right]f'\cdot f=0,\tag{4.27a}$$

$$(D_x^2-\mu D_x-\lambda)f'\cdot f=0.\tag{4.27b}$$

In order to show that the above equations *do* define a Bäcklund transformation, we need to investigate their compatibility condition. That is, we must show that there is no inconsistency between these equations. This compatibility is related to the inverse scattering method in a way that will be described later.

As will soon be demonstrated, the Bäcklund transformation given by (4.27) is related to (i) the inverse scattering formulation, (ii) the mKdV equation and (iii) the Miura transformation.

4.2.1 Inverse scattering formulation

The fundamental idea of the inverse scattering transformation is that the KdV equation,

$$u_t + 6uu_x + u_{xxx} = 0, \tag{4.28}$$

arises as the compatibility condition (4.30) of two linear differential operators (a Lax pair). We have

$$\begin{aligned} A\Psi &= 0, \\ L\Psi &= \lambda\Psi, \end{aligned} \tag{4.29}$$

where $A = \partial_t + (3\lambda + 3u)\partial_x + \partial_{xxx}$ and $L = \partial_{xx} + u$, and the compatibility condition is

$$[A, L] \equiv AL - LA = 0. \tag{4.30}$$

By setting $L_1 = A$ and $L_2 = L - \lambda$, this compatibility condition is equivalent to the commutation relation

$$[L_1, L_2] = 0. \tag{4.31}$$

We will prove that (4.31) is equivalent to the compatibility condition of the Bäcklund transformations (4.27). Since (4.27a) and (4.27b) may be thought of as linear differential equations for f or f', let us eliminate f or f' from these two equations. For this purpose, we introduce the rational dependent variable transformation

$$\begin{aligned} \Psi &= f'/f, \\ u &= 2(\log f)_{xx}. \end{aligned} \tag{4.32}$$

By employing formulae associated with the rational transformation,

$$\begin{aligned} (D_x f' \cdot f)/f^2 &= \Psi_x, \\ (D_x^2 f' \cdot f)/f^2 &= \Psi_{xx} + u\Psi, \\ (D_x^3 f' \cdot f)/f^2 &= \Psi_{xxx} + 3u\Psi_x, \end{aligned} \tag{4.33}$$

equations (4.27) are equivalent to

$$\begin{aligned} \Psi_t + (c_0 + 3\lambda)\Psi_x + \Psi_{xxx} + 3u\Psi_x &= 0, \\ \Psi_{xx} + u\Psi - \mu\Psi_x - \lambda\Psi &= 0. \end{aligned} \tag{4.34}$$

By introducing linear differential operators L_1, L_2 defined by

$$
\begin{aligned}
L_1 &= \partial_t + (c_0 + 3\lambda + 3u)\partial_x + \partial_x^3, \\
L_2 &= \partial_x^2 - \mu\partial_x + u - \lambda,
\end{aligned}
\tag{4.35}
$$

(4.34) may be rewritten as

$$
\begin{aligned}
L_1\Psi &= 0, \\
L_2\Psi &= 0.
\end{aligned}
\tag{4.36}
$$

The operators L_1 and L_2 satisfy a compatibility condition if their order may be interchanged. This means that

$$
L_1 L_2 \Psi = L_2 L_1 \Psi, \tag{4.37}
$$

or, equivalently,

$$
[L_1, L_2] = 0. \tag{4.38}
$$

In the case $\mu = c_0 = 0$, the linear differential operators L_1 and L_2 form the Lax pair (L_1, L_2) of the KdV equation [45]. As we have seen, the inverse scattering method and the direct method have a strong relation from the viewpoint of the Bäcklund transformation. However, the bilinear Bäcklund transformation not only gives rise to the original equation as its compatibility condition, as will be shown in Section 4.2.2 the bilinear Bäcklund transformation (4.27) also generates a new soliton equation.

4.2.2 The modified KdV (mKdV) equation

We start with the bilinear Bäcklund transformation formula,

$$
\begin{aligned}
\left[D_t + (c_0 + \lambda)D_x + D_x^3\right] f' \cdot f &= 0, \\
(D_x^2 - \mu D_x - \lambda) f' \cdot f &= 0.
\end{aligned}
\tag{4.39}
$$

Through logarithmic dependent variable transformations,

$$
\begin{aligned}
\phi &= \log(f'/f), \\
\rho &= \log(f'f),
\end{aligned}
\tag{4.40}
$$

and using the formulae (1.268) in Section 1.9, we obtain

$$
\phi_t + (c_0 + 3\lambda)\phi_x + \phi_{xxx} + 3\phi_x\rho_{xx} + \phi_x^3 = 0, \tag{4.41a}
$$

$$
\rho_{xx} + \phi_x^2 - \mu\phi_x - \lambda = 0. \tag{4.41b}
$$

Substituting ρ_{xx} from (4.41b) into (4.41a), we obtain

$$\phi_t + (c_0 + 3\lambda)\phi_x + \phi_{xxx} + 3\phi_x(\mu\phi_x + \lambda - \phi_x^2) + \phi_x^3 = 0. \tag{4.42}$$

Differentiating (4.42) with respect to x and putting $\phi_x = \widehat{v}$, we obtain the modified KdV equation (Gardner equation):

$$\widehat{v}_t + (c_0 + 6\lambda)\widehat{v}_x + \widehat{v}_{xxx} + 6\widehat{v}_x(-\widehat{v}^2 + \mu\widehat{v}) = 0. \tag{4.43}$$

In order to find solutions for the Gardner equation, we have only to solve the bilinear Bäcklund transformation formulae,

$$\begin{aligned} \left[D_t + (c_0 + \lambda)D_x + D_x^3\right] f' \cdot f &= 0, \\ (D_x^2 - \mu D_x - \lambda) f' \cdot f &= 0. \end{aligned} \tag{4.44}$$

4.2.3 The Miura transformation

In finding conserved quantities for the KdV equation,

$$u_t + 6uu_x + u_{xxx} = 0, \tag{4.45}$$

and the mKdV equation,

$$v_t + 24v^2 v_x + v_{xxx} = 0, \tag{4.46}$$

Miura [61] found that the solution u of the KdV equation is related to the solution v of the mKdV equation by the formula

$$u = (2v)^2 - 2iv_x. \tag{4.47}$$

Remarks

(1) Miura had very great difficulty in calculating higher-order conserved quantities of the KdV equation, and discovered the relation (4.47) by comparing the conserved quantities of the KdV equation with those of the mKdV equation. This was the beginning of discovery of the inverse scattering method for the KdV equation.

(2) The relation between $\widehat{v}$ and v is given by $\widehat{v} = 2iv$ when $c_0 = \mu = \lambda = 0$. □

Putting $\lambda = \mu = 0$ in the second formula in the bilinear Bäcklund transformation (4.39), we have

$$D_x^2 f' \cdot f = 0. \tag{4.48}$$

By using the formulae (1.268) of Section 1.9, the left-hand side of (4.48) may be rewritten as

$$
\begin{aligned}
(D_x^2 f' \cdot f)/(f'f) &= \left[\log(f'f)\right]_{xx} + \left[\log(f'/f)\right]_x^2 \\
&= 2(\log f)_{xx} + \left[\log(f'/f)\right]_{xx} + \left[\log(f'/f)\right]_x^2 .
\end{aligned}
$$

Because of the dependent variable transformation,

$$
\begin{aligned}
u &= 2(\log f)_{xx}, \\
v &= (1/2i)\left[\log(f'/f)\right]_x ,
\end{aligned}
\tag{4.49}
$$

the bilinear form $D_x^2 f' \cdot f = 0$ is simply the Miura transformation [61],

$$
u = (2v)^2 - 2iv_x. \tag{4.50}
$$

The procedure in which we exchanged the positions of f', f in

$$
P \equiv [F(D_t, D_x, D_y)f' \cdot f']f^2 - f'^2[F(D_t, D_x, D_y)f \cdot f] \tag{4.14 bis}
$$

to obtain the Bäcklund transformation

$$
\begin{aligned}
F_1(D_t, D_x, D_y)f' \cdot f &= 0, \\
F_2(D_t, D_x, D_y)f' \cdot f &= 0,
\end{aligned}
\tag{4.15 bis}
$$

can be applied to various soliton equations other than the KdV equation. Here, we list the Bäcklund transformations of the bilinear soliton equations listed in Section 1.8.

- The fifth-order KdV equation

$$
\left[D_x(D_t + D_x^5) - \frac{5}{6}D_s(D_s + D_x^3)\right] f \cdot f = 0, \tag{4.51}
$$

$$
\text{sub-condition :} \quad D_x(D_s + D_x^3)f \cdot f = 0. \tag{4.52}
$$

The Bäcklund transformation is

$$
(D_s + 3\lambda D_x + D_x^3)f' \cdot f = 0, \tag{4.53a}
$$

$$
D_x^2 f' \cdot f = \lambda f' f, \tag{4.53b}
$$

$$
(D_t + 15\lambda^2 D_x + D_x^5)f' \cdot f = 0. \tag{4.53c}
$$

Remark

The first and second equations of the above Bäcklund transformation are the same as those for the KdV equation. □

- Sawada–Kotera equation (1.254)

$$D_x(D_t + D_x^5) f \cdot f = 0. \tag{4.54}$$

The Bäcklund transformation is

$$D_x^3 f' \cdot f = \lambda f' f, \tag{4.55a}$$

$$\left(D_t - \frac{15}{2}\lambda D_x^2 - \frac{3}{2} D_x^5\right) f' \cdot f = 0. \tag{4.55b}$$

- Boussinesq equation (1.255)

$$(D_t^2 - D_x^2 - D_x^4) f \cdot f = 0. \tag{4.56}$$

The Bäcklund transformation is

$$(D_t + a D_x^2) f' \cdot f = 0, \tag{4.57a}$$

$$(a D_t D_x + D_x + D_x^3) f' \cdot f = 0, \tag{4.57b}$$

where $a^2 = -3$.

- Kadomtsev–Petviashvili (KP) equation (1.256)

$$(-4 D_x D_t + 3 D_y^2 + D_x^4) f \cdot f = 0. \tag{4.58}$$

The Bäcklund transformation is

$$(D_y - D_x^2) f' \cdot f = 0, \tag{4.59a}$$

$$(3 D_y D_x - 4 D_t + D_x^3) f' \cdot f = 0. \tag{4.59b}$$

- Model equation for shallow water waves (i)

$$\left[D_x(D_t - D_t D_x^2 + D_x) + \frac{1}{3} D_t (D_s + D_x^3)\right] f \cdot f = 0, \tag{4.60}$$

$$\text{sub-condition}: \quad D_x(D_s + D_x^3) f \cdot f = 0. \tag{4.61}$$

The Bäcklund transformation is

$$(D_s + 3\lambda D_x + D_x^3) f' \cdot f = 0, \tag{4.62a}$$

$$D_x^2 f' \cdot f = \lambda f' f + \mu D_x f' \cdot f, \tag{4.62b}$$

$$\left[(1 - 3\lambda) D_t - D_t D_x^2 + D_x\right] f' \cdot f = 0. \tag{4.62c}$$

Remark

The first and second equations in the above Bäcklund transformation are the same as those of the KdV equation. □

- Model equation for shallow water waves (ii)

$$D_x(D_t - D_t D_x^2 + D_x) f \cdot f = 0. \tag{4.63}$$

The Bäcklund transformation is

$$(D_x^3 - D_x) f' \cdot f = \lambda f' f, \tag{4.64a}$$

$$(3D_x D_t - 1) f' \cdot f = \mu D_x f' \cdot f. \tag{4.64b}$$

- Toda lattice equation (1.260)

$$\left[D_t^2 - 4\sinh^2\left(\frac{1}{2}D_n\right)\right] f \cdot f = 0. \tag{4.65}$$

The Bäcklund transformations are

$$\left[D_t \exp\left(-\frac{D_n}{2}\right) - 2\lambda \sinh\left(\frac{D_n}{2}\right)\right] f' \cdot f = 0, \tag{4.66a}$$

$$\left[D_t + \lambda^{-1}(\exp(-D_n) - 1)\right] f' \cdot f = 0, \tag{4.66b}$$

and

$$D_t f' \cdot f + 2\alpha \sinh\left(\frac{D_n}{2}\right) g' \cdot g = 0, \tag{4.67a}$$

$$D_t g' \cdot g + 2\alpha^{-1} \sinh\left(\frac{D_n}{2}\right) f' \cdot f = 0, \tag{4.67b}$$

$$\left[\beta_1 \sinh\left(\frac{D_n}{2}\right) + \cosh\left(\frac{D_n}{2}\right)\right] g' \cdot g = f' f, \tag{4.67c}$$

$$\left[\beta_2 \sinh\left(\frac{D_n}{2}\right) + \cosh\left(\frac{D_n}{2}\right)\right] f' \cdot f = g' g, \tag{4.67d}$$

$$\text{condition :} \quad \alpha^{-1}(\beta_1^2 - 1) = \alpha(\beta_2^2 - 1). \tag{4.68}$$

4.3 The Bäcklund transformation for the KP equation

Let us investigate the bilinear Bäcklund transformation formulae for the KP equation. This equation is

$$(-4D_x D_t + 3D_y^2 + D_x^4) f \cdot f = 0, \tag{4.58 bis}$$

and its Bäcklund transformation is

$$(D_y - D_x^2) f' \cdot f = 0, \qquad (4.59\text{a } bis)$$

$$(3D_y D_x - 4D_t + D_x^3) f' \cdot f = 0. \qquad (4.59\text{b } bis)$$

Remark

The above Bäcklund transformation formulae coincide with the first modified KP equation [15] if we set $D_1 = D_x, D_2 = D_y, D_3 = D_z$. □

We have already explained that the solution for the KP equation τ may be expressed in terms of wronskian and grammian determinants. Let us investigate how τ' is expressed. To start with the conclusions, there are two kinds of solution τ': (i) a solution with the same number of solitons but with different phase, and (ii) a solution with the number of solitons increased by one. In these two cases, (4.59) is simply a Plücker relation and a pfaffian identity, respectively.

In this section, we give various expressions for τ and τ' and confirm that they satisfy the first equation in the Bäcklund transformations given by $(D_1^2 + D_2)\tau \cdot \tau' = 0$.

4.3.1 Wronskian expression

We first define $\tau_N^{(n)}$ by

$$\tau_N^{(n)} \equiv \begin{vmatrix} f_1^{(n)} & f_1^{(n+1)} & \cdots & f_1^{(n+N-1)} \\ f_2^{(n)} & f_2^{(n+1)} & \cdots & f_2^{(n+N-1)} \\ \vdots & \vdots & & \vdots \\ f_N^{(n)} & f_N^{(n+1)} & \cdots & f_N^{(n+N-1)} \end{vmatrix}$$
$$= (d_n, d_{n+1}, \ldots, d_{n+N-1}, N, \ldots, 2, 1), \qquad (4.69)$$

where the elements in the above pfaffian are defined by

$$(d_m, j) = \frac{\partial}{\partial x_m} f_j = \frac{\partial^m}{\partial x^m} f_j \quad (j = 1, 2, \ldots, N),$$
$$(d_m, d_n) = 0 \quad (m, n = 0, 1, 2, \ldots). \qquad (4.70)$$

Remarks

(1) The suffix N $(= 0, 1, 2, \ldots)$ in $\tau_N^{(n)}$ represents the size of determinant, that is, the number of solitons, and n $(= \ldots, -1, 0, 1, 2, \ldots)$ is a parameter introduced to refer to different choices of τ_N.

(2) The wronskian $\tau_N^{(n)}$ is expressed by means of a Maya diagram as

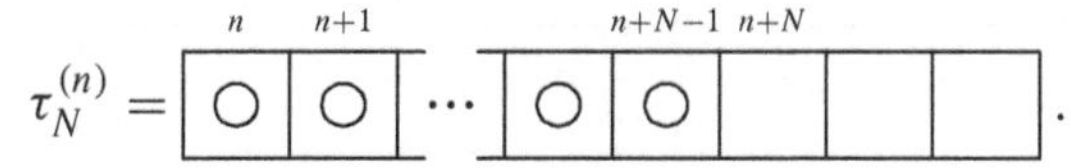

□

There are two choices for τ, τ' expressed in terms of $\tau_N^{(n)}$ [62]

$$\text{(i)}\ \ \tau = \tau_N^{(0)}, \quad \tau' = \tau_N^{(1)},$$
$$\text{(ii)}\ \ \tau = \tau_N^{(0)}, \quad \tau' = \tau_{N+1}^{(0)}.$$

In case (i), N, the number of functions f_j contained in τ, τ' (or equivalently, the number of solitons) is the same, but the phase constants δ_i, which show the position of the solitons, change.

Remark

In the case of a two-soliton solution, τ, τ' are written as

$$\begin{aligned}
\tau &\propto 1 + \exp(\eta_1) + \exp(\eta_2) + c_{12}\exp(\eta_1 + \eta_2),\\
\tau' &\propto 1 + \exp(\eta_1 + \delta_1) + \exp(\eta_2 + \delta_2)\\
&\quad + c_{12}\exp(\eta_1 + \eta_2 + \delta_1 + \delta_2),
\end{aligned}$$

where

$$\eta_i = \xi_i - \widehat{\xi}_i, \quad \xi_i = \sum_n p_i^n x_n, \quad \widehat{\xi}_i = \sum_n q_i^n x_n$$

and

$$\exp(\delta_i) = \frac{p_i}{q_i}, \quad c_{12} = \frac{(p_1 - p_2)(q_1 - q_2)}{(p_1 - q_2)(q_1 - p_2)}.$$

□

In order to confirm that τ, τ' satisfy the Bäcklund transformation formula (4.59a), we determine the derivatives of $\tau' = \tau_N^{(1)}$, and write the results in terms of Maya diagrams:

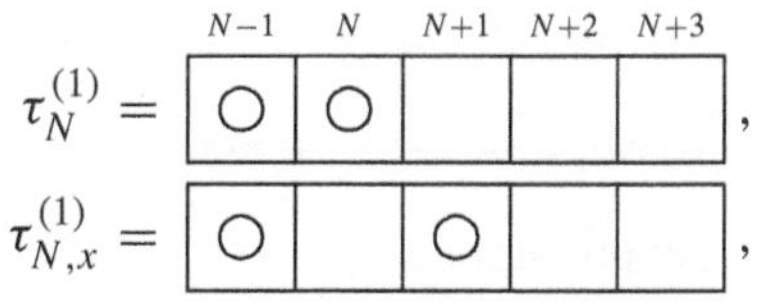

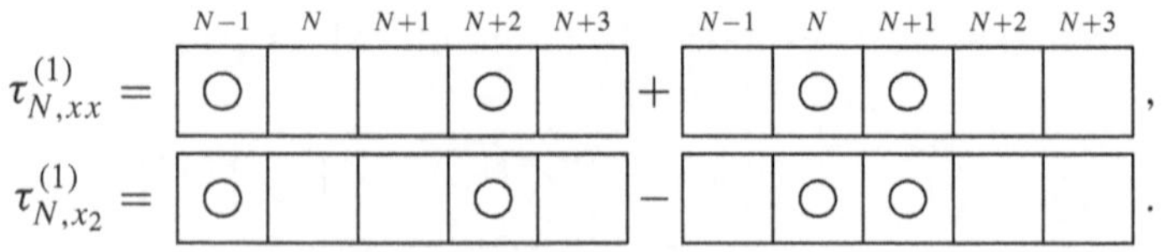

The derivatives of $\tau = \tau_N^{(0)}$ are obtained by reducing the number on each cell in the Maya diagrams by one. Substituting these results into the left-hand side of (4.59a), that is

$$(\tau_{xx} + \tau_{x_2})\tau' - 2\tau_x\tau'_x + \tau(\tau'_{xx} - \tau'_{x_2}),$$

we obtain the Maya diagram expression

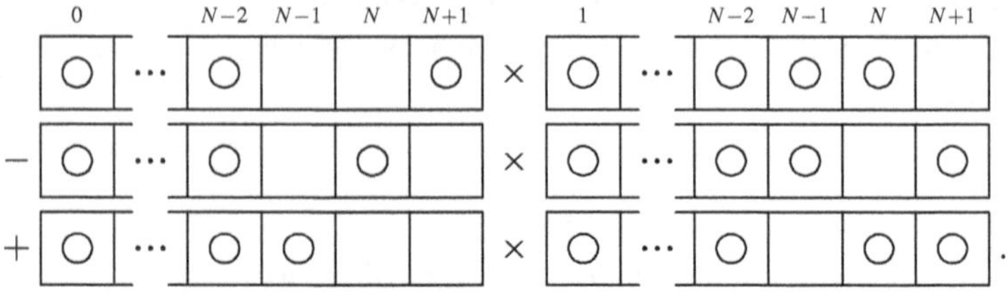

Neglecting cells common to every expression and exchanging positions of τ and τ', we finally obtain

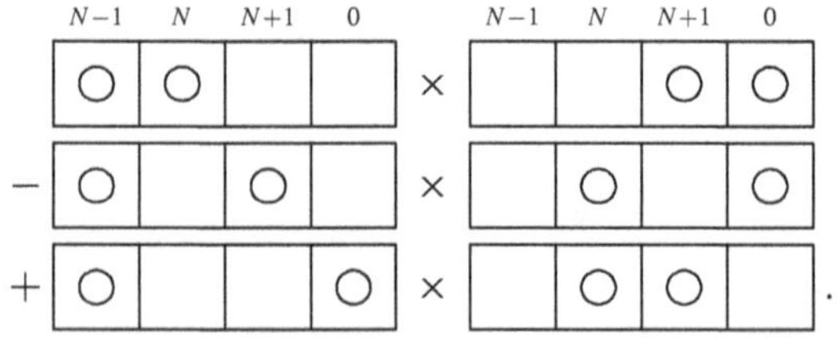

Since this is nothing but a Plücker relation, it is identically zero. Hence, we have shown that τ, τ' satisfy the first of the Bäcklund transformation equations (4.59a).

In case (ii), τ' contains one more function, f_{N+1}, than τ. Since the formulae for the derivatives of τ',

$$\tau' = \tau_{N+1}^{(0)} = (d_0, d_1, \ldots, d_{N-1}, d_N, N+1, N, \ldots, 2, 1),$$

are obtained by replacing N with $N+1$ in those for $\tau = \tau_N^{(0)}$, substitution of the above results into the left-hand side of (4.59a), that is

$$(\tau_{xx} + \tau_{x_2})\tau' - 2\tau_x\tau'_x + \tau(\tau'_{xx} - \tau'_{x_2}),$$

yields the Maya diagram expression

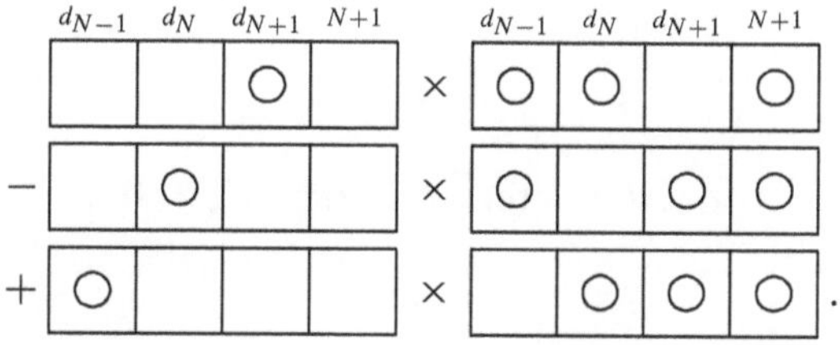

Using (2.96′), this may be rewritten as

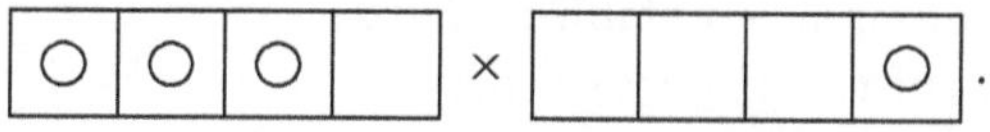

However, the above equation is identically equal to zero because the number of the labels corresponding to taking derivatives d_m and the number representing solitons j are not the same. Therefore, we have proved that τ, τ' satisfy the Bäcklund transformation formula (4.59a) [38].

4.3.2 Grammian expression

We first define $\tau_N^{[n]}$ by

$$\begin{aligned} \tau_N^{[n]} &\equiv \det((m_{ij})_n)_{1\le i,j\le N}, \\ (m_{ij})_n &\equiv c_{ij} + (-)^n \int^x f_i^{(n)} g_j^{(-n)}\,\mathrm{d}x, \end{aligned} \tag{4.71}$$

where f_j and g_j satisfy

$$\begin{aligned} \frac{\partial}{\partial x_m} f_j &= \frac{\partial^m}{\partial x^m} f_j \equiv f_j^{(m)}, \\ \frac{\partial}{\partial x_m} g_j &= (-)^{m-1} \frac{\partial^m}{\partial x^m} g_j \equiv (-)^{m-1} g_j^{(m)}, \end{aligned} \tag{4.72}$$

for $j = 1, 2, \ldots, N$ and $m = 1, 2, \ldots$. Using the above expressions, τ, τ' are written in two different ways.

(i) First as

$$\tau = \tau_N^{[0]}, \quad \tau' = \tau_N^{[1]}. \tag{4.73}$$

In this case, τ, τ' have the same number of functions but different phases.

Each element of $\tau' = \tau_N^{[1]}$, given by $(m_{ij})_1$, may be rewritten, using integration by parts, as

$$\begin{aligned} (m_{ij})_1 &= c_{ij} + \int^x f_i^{(0)} g_j^{(0)}\,\mathrm{d}x - f_i^{(0)} g_j^{(-1)} \\ &= (m_{ij})_0 - (d_{-1}, d_0^*, i, j^*), \end{aligned} \tag{4.74}$$

where we have

$$
\begin{aligned}
&(d_m, j^*) = g_j^{(m)}, \quad (d_m^*, j) = f_j^{(m)}, \\
&(d_m, d_n^*) = (d_m, j) = (d_m^*, j^*) = 0.
\end{aligned} \tag{4.75}
$$

Therefore,

$$
\tau_N^{[1]} = \tau_N^{[0]} - (d_{-1}, d_0^*, 1, 2, \ldots, N, N^*, \ldots, 2^*, 1^*), \tag{4.76}
$$

using the addition formula for pfaffians (2.102).

The derivatives of $\tau = \tau_n^{[0]}$ are given by

$$
\tau = (1, 2, \ldots, N, N^*, \ldots, 2^*, 1^*) \equiv (\bullet), \tag{4.77a}
$$

$$
\tau_x = (d_0, d_0^*, \bullet), \tag{4.77b}
$$

$$
\tau_{xx} = (d_1, d_0^*, \bullet) + (d_0, d_1^*, \bullet), \tag{4.77c}
$$

$$
\tau_{x_2} = -(d_1, d_0^*, \bullet) + (d_0, d_1^*, \bullet), \tag{4.77d}
$$

$$
\tau' = \tau - (d_{-1}, d_0^*, \bullet), \tag{4.77e}
$$

$$
\tau'_x = (d_{-1}, d_1^*, \bullet), \tag{4.77f}
$$

$$
\tau'_{xx} = -(d_0, d_1^*, \bullet) - (d_{-1}, d_2^*, \bullet) - (d_{-1}, d_1^*, d_0, d_0^*, \bullet), \tag{4.77g}
$$

$$
\tau'_{x_2} = (d_0, d_1^*, \bullet) - (d_{-1}, d_2^*, \bullet) + (d_{-1}, d_1^*, d_0, d_0^*, \bullet). \tag{4.77h}
$$

Substituting the above results into the left-hand side of the first equation of the Bäcklund transformation,

$$
(\tau_{xx} + \tau_{x_2})\tau' - 2\tau_x \tau'_x + \tau(\tau'_{xx} - \tau'_{x_2}),
$$

gives the Maya diagram expression

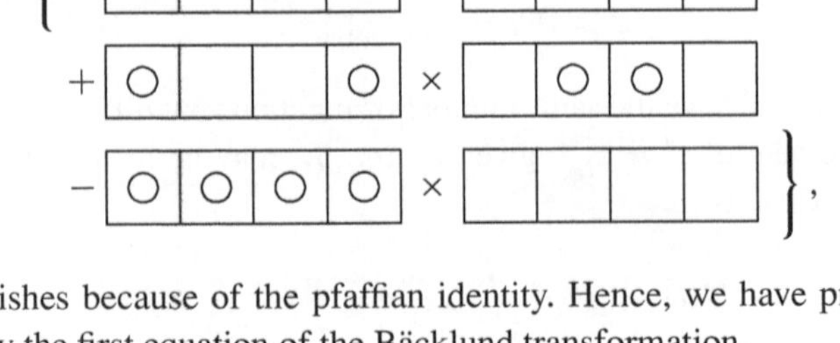

which vanishes because of the pfaffian identity. Hence, we have proved that τ, τ' satisfy the first equation of the Bäcklund transformation.

(ii) Secondly, as

$$\tau = (1, 2, \ldots, N, N^*, \ldots, 2^*, 1^*) \equiv (\bullet), \tag{4.78a}$$

$$\tau' = (1, 2, \ldots, N, N+1, d_0^*, N^*, \ldots, 2^*, 1^*) \equiv -(d_0^*, N+1, \bullet). \tag{4.78b}$$

In this case, τ' contains one more function $(d_0, N+1)$ compared with τ. Its derivatives are

$$\tau'_x = -(d_1^*, N+1, \bullet), \tag{4.79a}$$

$$\tau'_{xx} = -(d_2^*, N+1, \bullet) - (d_0, d_0^*, d_1^*, N+1, \bullet), \tag{4.79b}$$

$$\tau'_{x_2} = -(d_2^*, N+1, \bullet) + (d_0, d_0^*, d_1^*, N+1, \bullet). \tag{4.79c}$$

Substitution of the above results into the left-hand side of the first equation of the Bäcklund transformation,

$$(\tau_{xx} + \tau_{x_2})\tau' - 2\tau_x\tau'_x + \tau(\tau'_{xx} - \tau'_{x_2}),$$

gives the Maya diagram expression

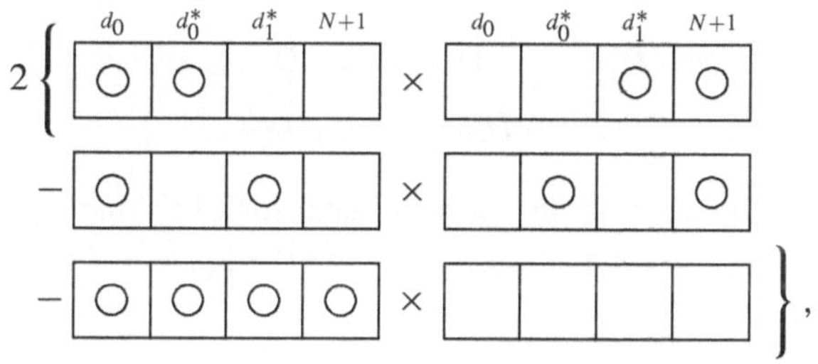

which vanishes because of the pfaffian identity. Hence, we have proved that τ, τ' solve the first equation of the Bäcklund transformation given by (4.59a).

Following the same procedure, we can prove that τ, τ' solve the other equation (4.59b).

4.4 The Bäcklund transformation for the BKP equation

In this section, we find a Bäcklund transformation for the BKP equation [46],

$$\left[(D_3 - D_1^3)D_{-1} + 3D_1^2\right]\tau \cdot \tau = 0. \tag{4.80}$$

We define τ' to be another solution of (4.80) and consider

$$\begin{aligned} P \equiv\ & \{[(D_3 - D_1^3)D_{-1} + 3D_1^2]\tau \cdot \tau\}\tau'\tau' \\ & - \tau\tau\{[(D_3 - D_1^3)D_{-1} + 3D_1^2]\tau' \cdot \tau'\}. \end{aligned} \tag{4.81}$$

Using the exchange formulae,

$$\left[D_3 D_{-1}\tau \cdot \tau\right]\tau'\tau' - \tau\tau\left[D_3 D_{-1}\tau' \cdot \tau'\right] = 2D_{-1}\left[D_3\tau \cdot \tau'\right] \cdot \tau'\tau, \tag{4.82a}$$

$$\left[D_1^2\tau \cdot \tau\right]\tau'\tau' - \tau\tau\left[D_1^2\tau' \cdot \tau'\right] = 2D_1\left[D_1\tau \cdot \tau'\right] \cdot \tau'\tau, \tag{4.82b}$$

$$\begin{aligned} & \left[D_1^3 D_{-1}\tau \cdot \tau\right]\tau'\tau' - \tau\tau\left[D_1^3 D_{-1}\tau' \cdot \tau'\right] \\ & = 2D_{-1}\left[D_1^3\tau \cdot \tau'\right] \cdot \tau'\tau + 6D_1(D_1 D_{-1}\tau \cdot \tau') \cdot (D_1\tau' \cdot \tau), \end{aligned} \tag{4.82c}$$

we obtain

$$\begin{aligned} P =\ & 2D_{-1}[(D_3 - D_1^3)\tau \cdot \tau'] \cdot \tau'\tau \\ & - 6D_1[(D_1 D_{-1} - 1)\tau \cdot \tau'] \cdot (D_1\tau' \cdot \tau). \end{aligned} \tag{4.83}$$

From the above,

$$(D_1 D_{-1} - 1)\tau \cdot \tau' = \lambda D_1\tau' \cdot \tau, \tag{4.84a}$$

$$(D_1^3 - D_3)\tau \cdot \tau' = \mu\tau\tau', \tag{4.84b}$$

where λ, μ, which are constants, are candidates for a Bäcklund transformation. We call the above equations the modified BKP equation. We may derive the following results.

4.4.1 Inverse scattering form

Let us make the rational dependent variable transformation

$$\Psi = \tau/\tau', \quad w = 2(\log\tau')_x, \tag{4.85}$$

and introduce linear differential operators L_1, L_2 given by

$$L_1 = \partial_x\partial_y + w_y - 1 + \lambda\partial_x, \tag{4.86}$$

$$L_2 = \partial_x^3 - \partial_t + 3w_x\partial_x - \mu. \tag{4.87}$$

Then (4.84a) and (4.84b) may be rewritten as

$$L_1\Psi = 0, \quad L_2\Psi = 0. \tag{4.88}$$

From the compatibility condition $L_1L_2\Psi = L_2L_1\Psi$, we obtain the BKP equation,

$$w_{yt} - w_{xxxy} - 3(w_x w_y)_x + 3w_{xx} = 0, \tag{4.89}$$

where we have put $x_1 = x, x_{-1} = y, x_3 = t$.

4.4.2 The modified BKP equation

Through the logarithmic dependent variable transformation

$$\phi = \log(\tau/\tau'), \quad \rho = \log(\tau\tau'), \tag{4.90}$$

(4.84a) and (4.84b) may be rewritten as

$$\begin{aligned} \rho_{xy} + \phi_x\phi_y + \lambda\phi_x - 1 = 0, \\ \phi_t - \phi_{xxx} - 3\phi_x\rho_{xx} - \phi_x^3 + \mu = 0. \end{aligned} \tag{4.91}$$

4.5 The solution of the modified BKP equation

We learned in Chapter 3 that the BKP equation has solutions expressed in terms of pfaffians. In this section, we describe solutions τ, τ' of the modified BKP equation, (4.84a), (4.84b), with parameters $\lambda = \mu = 0$,

$$(D_1D_{-1} - 1)\tau \cdot \tau' = 0, \tag{4.92a}$$

$$(D_1^3 - D_3)\tau \cdot \tau' = 0. \tag{4.92b}$$

We have seen that τ may be expressed as the pfaffian

$$\begin{aligned} &\tau = (1, 2, \dots, 2N), \\ &(i, j) = c_{ij} + \int^x D_x f_i(x) \cdot f_j(x)\, \mathrm{d}x \quad (x = x_1), \end{aligned} \tag{4.93}$$

where $f_i(x)$ for $i = 1, 2, \dots, 2N$ satisfy the linear differential equations

$$\frac{\partial}{\partial x_n} f_i(x) = \frac{\partial^n}{\partial x^n} f_i(x) \quad (n = -1, 1, 3, \dots). \tag{4.94}$$

In the case $n = -1$, (4.94) is

$$\frac{\partial}{\partial x_{-1}} f_i(x) = \int^x f_i(x)\, \mathrm{d}x. \tag{4.95}$$

Then τ' is given by the pfaffian

$$\begin{aligned}\tau' &= (d_0, 1, 2, \ldots, 2N, 2N+1) \\ &= (d_0, 2N+1, 1, 2, \ldots, 2N),\end{aligned} \tag{4.96}$$

where

$$(d_n, i) = \frac{\partial^n}{\partial x^n} f_i, \quad (d_m, d_n) = 0 \qquad (m, n = -1, 0, 1, 2, \ldots). \tag{4.97}$$

In order to confirm that τ, τ' satisfy (4.92a), (4.92b), we calculate the derivatives of τ':

$$\tau' = (d_0, 2N+1, 1, 2, \ldots, 2N) \equiv (d_0, 2N+1, \bullet), \tag{4.98a}$$

$$\frac{\partial}{\partial x_{-1}}\tau' = (d_{-1}, 2N+1, \bullet), \tag{4.98b}$$

$$\frac{\partial^2}{\partial x \partial x_{-1}}\tau' = (d_0, 2N+1, \bullet) + (d_{-1}, d_0, d_1, 2N+1, \bullet), \tag{4.98c}$$

$$\frac{\partial}{\partial x}\tau' = (d_1, 2N+1, \bullet), \tag{4.98d}$$

$$\frac{\partial^3}{\partial x^3}\tau' = (d_3, 2N+1, \bullet) + (d_0, d_1, d_2, 2N+1, \bullet), \tag{4.98e}$$

$$\frac{\partial}{\partial x_3}\tau' = (d_3, 2N+1, \bullet) - (d_0, d_1, d_2, 2N+1, \bullet). \tag{4.98f}$$

Substitution of these expressions and (3.78) into (4.92a), (4.92b) gives the Maya diagram expressions

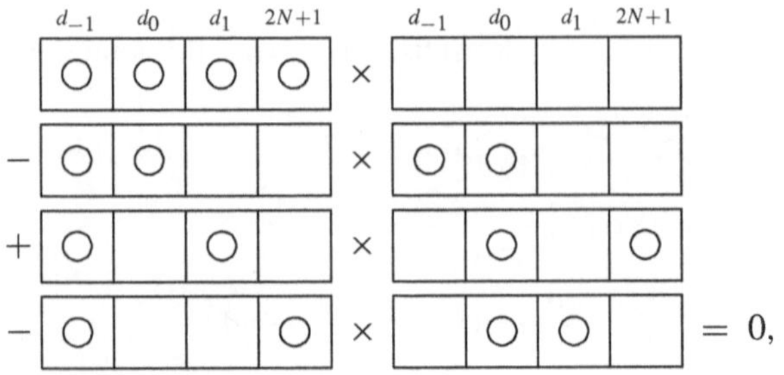

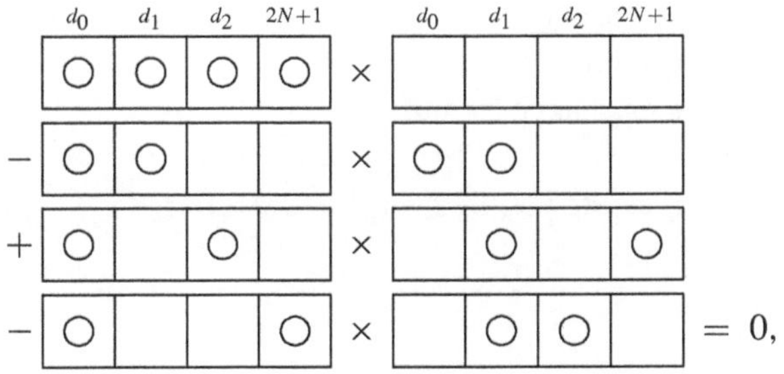

which are simply pfaffian identities. Therefore, we have confirmed that τ, τ' satisfy (4.92a), (4.92b).

4.6 The Bäcklund transformation for the two-dimensional Toda equation

We consider the two-dimensional Toda equation

$$\frac{\partial^2}{\partial s \partial x} \log(\varepsilon + V_n) = V_{n+1} - 2V_n + V_{n-1}, \quad \varepsilon = 0, 1, \tag{4.99}$$

where $\varepsilon = 1$ and $\varepsilon = 0$ correspond to (i) the Toda lattice equation and (ii) the Toda molecule equation, respectively.

Through the dependent variable transformation

$$V_n \equiv (\log \tau_n)_{xs},$$

(4.99) are integrated to give the following bilinear forms: (i) the two-dimensional Toda *lattice* equation ($n = \ldots, -1, 0, 1, \ldots$),

$$D_x D_s \tau_n \cdot \tau_n = 2(\tau_{n+1}\tau_{n-1} - \tau_n^2), \tag{4.100a}$$

and (ii) the two-dimensional Toda *molecule* equation ($n = 1, 2, \ldots, N$),

$$D_x D_s \tau_n \cdot \tau_n = 2\tau_{n+1}\tau_{n-1}, \tag{4.100b}$$

with boundary conditions $V_0 = V_{N+1} = 0$.

Let us find a Bäcklund transformation which connects the solution τ_n of the Toda equation with another solution τ_n' [53] by considering

$$\begin{aligned} P = &\left[D_x D_s \tau_n \cdot \tau_n - 2\tau_{n+1}\tau_{n-1} + 2\varepsilon\tau_n^2\right]\tau_n'^2 \\ &- \tau_n^2\left[D_x D_s \tau_n' \cdot \tau_n' - 2\tau_{n+1}'\tau_{n-1}' + 2\varepsilon\tau_n'^2\right]. \end{aligned} \tag{4.101}$$

If $P = 0$, then $\tau_n \neq 0$ satisfies the Toda equation if and only if $\tau'_n \neq 0$ also satisfies the same equation.

By employing the exchange formula

$$[D_x D_s \tau_n \cdot \tau_n]\, \tau_n'^2 - \tau_n^2 \left[D_x D_s \tau'_n \cdot \tau'_n\right] = 2D_x(D_s \tau_n \cdot \tau'_n) \cdot \tau'_n \tau_n \quad (4.102)$$

and the identity

$$D_x \tau_{n+1}\tau'_{n-1} \cdot \tau'_n \tau_n = (D_x \tau_{n+1} \cdot \tau'_n)\tau'_{n-1}\tau_n + \tau_{n+1}\tau'_n(D_x \tau'_{n-1} \cdot \tau_n), \quad (4.103)$$

P may be rewritten as

$$\begin{aligned} \tfrac{1}{2}P &= D_x\left[D_s \tau_n \cdot \tau'_n - \lambda \tau_{n+1}\tau'_{n-1}\right] \cdot \tau'_n \tau_n \\ &\quad + \lambda\left[D_x \tau_{n+1} \cdot \tau'_n + \lambda^{-1}\tau_n \tau'_{n+1}\right]\tau'_{n-1}\tau_n \\ &\quad - \lambda\left[D_x \tau_n \cdot \tau'_{n-1} + \lambda^{-1}\tau_{n-1}\tau'_n\right]\tau'_n \tau_{n+1}, \end{aligned} \quad (4.101')$$

where λ is a free parameter. Therefore, $P = 0$ if the bilinear equations

$$D_s \tau_n \cdot \tau'_n = \lambda \tau_{n+1}\tau'_{n-1} - \mu \tau_n \tau'_n, \quad (4.104a)$$

$$D_x \tau_{n+1} \cdot \tau'_n = -\lambda^{-1}\tau_n \tau'_{n+1} + \nu \tau_{n+1}\tau'_n, \quad (4.104b)$$

(μ, ν are also free parameters) are satisfied for all n. These bilinear equations give the Bäcklund transformation connecting the solutions τ_n, τ'_n of the Toda equation. The Bäcklund transformation equations (4.104a), (4.104b) may be rewritten in several ways [53].

4.6.1 Lax pair

We introduce new dependent variables Ψ_n, V_n, I_n by means of

$$\tau'_n \equiv \Psi_n \tau_n, \quad (4.105)$$

$$I_n \equiv \frac{\partial}{\partial x}\log(\tau_n/\tau_{n+1}) = \frac{D_x \tau_n \cdot \tau_{n+1}}{\tau_n \tau_{n+1}}, \quad (4.106)$$

$$V_n \equiv (\log \tau_n)_{xs} = \frac{\tau_{n+1}\tau_{n-1}}{\tau_n^2} - \varepsilon. \quad (4.107)$$

Then (4.104a) and (4.104b) are transformed into

$$\frac{\partial \Psi_n}{\partial s} = -\lambda(V_n + \varepsilon)\Psi_{n-1} + \mu\Psi_n, \tag{4.108a}$$

$$\frac{\partial \Psi_n}{\partial x} = \lambda^{-1}\Psi_{n+1} - \nu\Psi_n - I_n\Psi_n. \tag{4.108b}$$

Introducing linear differential-difference operators L_1, L_2 defined by

$$L_1 \equiv \frac{\partial}{\partial s} + \lambda(V_n + \varepsilon)\exp\left(-\frac{\partial}{\partial n}\right) - \mu, \tag{4.109a}$$

$$L_2 \equiv \frac{\partial}{\partial x} - \lambda^{-1}\exp\left(\frac{\partial}{\partial n}\right) + I_n + \nu, \tag{4.109b}$$

we have, from the compatibility condition

$$L_1 L_2 \Psi = L_2 L_1 \Psi, \tag{4.110}$$

a set of partial differential equations for I_n, V_n,

$$\begin{aligned} \frac{\partial}{\partial x}\log(\varepsilon + V_n) &= I_{n-1} - I_n, \\ \frac{\partial I_n}{\partial s} &= V_n - V_{n+1}. \end{aligned} \tag{4.111}$$

Elimination of I_n from (4.111) gives the Toda equation, and therefore (4.109a), (4.109b) are its Lax pair.

4.6.2 The modified Toda equation

We define new dependent variables u_n by

$$u_n \equiv \frac{\partial}{\partial s}\log\left(\frac{\tau_n}{\tau_n'}\right) = \frac{D_s \tau_n \cdot \tau_n'}{\tau_n \tau_n'}, \tag{4.112}$$

and v_n by

$$v_n \equiv -\frac{\partial}{\partial x}\log\left(\frac{\tau_n}{\tau_{n-1}'}\right) = -\frac{D_x \tau_n \cdot \tau_{n-1}'}{\tau_n \tau_{n-1}'}. \tag{4.113}$$

Using (4.104) we may rewrite these as

$$u_n = \frac{\lambda \tau_{n+1}\tau_{n-1}'}{\tau_n \tau_n'} - \mu \tag{4.112$'$}$$

and

$$v_n = \frac{\lambda^{-1}\tau_{n-1}\tau_n'}{\tau_n\tau_{n-1}'} - \nu. \tag{4.113'}$$

Differentiating $\log(\mu + u_n)$ and $\log(\nu + v_n)$ with respect to x and s, respectively, we obtain

$$\begin{aligned}\frac{\partial}{\partial x}\log(\mu + u_n) &= \frac{\partial}{\partial x}\left[-\log\left(\frac{\tau_n}{\tau_{n-1}'}\right) + \log\left(\frac{\tau_{n+1}}{\tau_n'}\right)\right]\\ &= v_n - v_{n+1}, \qquad (4.114a)\\ \frac{\partial}{\partial s}\log(\nu + v_n) &= \frac{\partial}{\partial s}\left[\log\left(\frac{\tau_{n-1}}{\tau_{n-1}'}\right) - \log\left(\frac{\tau_n}{\tau_n'}\right)\right]\\ &= u_{n-1} - u_n, \qquad (4.114b)\end{aligned}$$

which are equivalent to

$$\frac{\partial}{\partial x}u_n = (\mu + u_n)(v_n - v_{n+1}), \tag{4.115a}$$

$$\frac{\partial}{\partial s}v_n = (\nu + v_n)(u_{n-1} - u_n). \tag{4.115b}$$

We call the above system the *modified Toda equations*.

Remark

Let the independent variable n take the finite number of values $n = 1, 2, \ldots, N$. Imposing the boundary conditions $u_0 = v_{N+1} = 0$ and putting parameters $\mu = \nu = 0$, (4.115a), (4.115b) give

$$\frac{\partial}{\partial x}u_n = u_n(v_n - v_{n+1}), \tag{4.116a}$$

$$\frac{\partial}{\partial s}v_n = v_n(u_{n-1} - u_n). \tag{4.116b}$$

This coupled system of differential equations describes the interaction of $2N$ waves u_n, v_n for $n = 1, 2, \ldots, 2N$ [53]. If $N = 1$, (4.116a), (4.116b) reduce to the two-wave interaction equation (1.78),

$$\begin{aligned}\frac{\partial}{\partial x}u_1 &= u_1 v_1,\\ \frac{\partial}{\partial s}v_1 &= -v_1 u_1.\end{aligned} \tag{4.117}$$

□

4.6.3 Miura transformation

The solutions V_n of the Toda equation and u_n, v_n of the modified Toda equations are expressed as follows:

$$\varepsilon + V_n = \frac{\tau_{n+1}\tau_{n-1}}{\tau_n^2}, \tag{4.118a}$$

$$\mu + u_n = \lambda \frac{\tau_{n+1}\tau'_{n-1}}{\tau_n \tau'_n}, \tag{4.118b}$$

$$\nu + v_n = \lambda^{-1} \frac{\tau_{n-1}\tau'_n}{\tau_n \tau'_{n-1}}. \tag{4.118c}$$

Comparing these, we see that V_n may be expressed in terms of u_n, v_n as follows:

$$\varepsilon + V_n = (\mu + u_n)(\nu + v_n). \tag{4.119}$$

Substituting the above expression into the left-hand side of the Toda equation,

$$\frac{\partial^2}{\partial x \partial s} \log(\varepsilon + V_n) = V_{n+1} - 2V_n + V_{n-1}, \tag{4.120}$$

and considering the fact that u_n, v_n satisfy the modified Toda equations (4.115), we obtain

$$\begin{aligned}
&\frac{\partial}{\partial s}\left[\frac{\partial}{\partial x}\log(\mu + u_n)\right] + \frac{\partial}{\partial x}\left[\frac{\partial}{\partial s}\log(\nu + v_n)\right] \\
&= \frac{\partial}{\partial s}[v_n - v_{n+1}] + \frac{\partial}{\partial x}[u_{n-1} - u_n] \\
&= (\nu + v_n)(u_{n-1} - u_n) - (\nu + v_{n+1})(u_n - u_{n-1}) \\
&\quad + (\mu + u_{n-1})(v_{n-1} - v_n) - (\mu + u_n)(v_n - v_{n-1}) \\
&= (\mu + u_{n+1})(\nu + v_{n+1}) - 2(\mu + u_n)(\nu + v_n) \\
&\quad + (\mu + u_{n-1})(\nu + v_{n-1}) \\
&= V_{n+1} - 2V_n + V_{n-1},
\end{aligned} \tag{4.121}$$

which is simply the right-hand side of the Toda equation. Therefore, if u_n, v_n satisfy the modified Toda equation, then V_n satisfies the Toda equation. This shows that (4.119) is a Miura transformation connecting the solutions of nonlinear differential equations.

4.7 Solutions of the two-dimensional modified Toda equation

As we have explained in Chapter 3, solutions τ_n to the Toda equation,

$$D_x D_s \tau_n \cdot \tau_n = 2(\tau_{n+1}\tau_{n-1} - \varepsilon \tau_n^2), \tag{4.122}$$

have different forms for (a) the Toda lattice equation ($\varepsilon = 1, n = \ldots, -1, 0, 1, \ldots$) and (b) the Toda molecule equation ($\varepsilon = 0, n = 1, 2, \ldots, N$). Therefore, the expressions for solutions τ_n' of the modified Toda equation,

$$\begin{aligned} D_s \tau_n \cdot \tau_n' &= \lambda \tau_{n+1} \tau_{n-1}' - \mu \tau_n \tau_n', \\ D_x \tau_{n+1} \cdot \tau_n' &= -\lambda^{-1} \tau_n \tau_{n+1}' + \nu \tau_{n+1} \tau_n', \end{aligned} \tag{4.123}$$

where λ, μ and ν are free parameters, are also different for these two cases.

4.7.1 Structure of the Bäcklund transformation for the Toda lattice equation

We have shown that the N-soliton solution τ_n of the Toda lattice equation can be expressed as an Nth-order wronskian

$$\begin{aligned} \tau_n &= \begin{vmatrix} \phi_1(n) & \phi_1(n+1) & \cdots & \phi_1(n+N-1) \\ \phi_2(n) & \phi_2(n+1) & \cdots & \phi_2(n+N-1) \\ \vdots & \vdots & & \vdots \\ \phi_N(n) & \phi_N(n+1) & \cdots & \phi_N(n+N-1) \end{vmatrix} \\ &\equiv [\phi(n), \phi(n+1), \ldots, \phi(n+N-1)] \\ &= (d_0, d_1, \ldots, d_{N-1}, N, \ldots, 2, 1), \end{aligned} \tag{4.124}$$

where each $\phi_i(n)$ satisfies

$$\begin{aligned} \frac{\partial \phi_i(n)}{\partial x} &= \phi_i(n+1), \\ \frac{\partial \phi_i(n)}{\partial s} &= -\phi_i(n-1). \end{aligned} \tag{4.125}$$

As for the KP equation, there are two different kinds of solutions τ_n' obtained from τ_n through the Bäcklund transformation.

Solution with the same number of solitons but a different phase
Here we choose $\mu = \lambda = -b/a$ and $\nu = \lambda^{-1}$. The Bäcklund transformation formulae are given by

$$D_s \tau_n \cdot \tau'_n = -\frac{b}{a}(\tau_{n+1}\tau'_{n-1} - \tau_n \tau'_n), \tag{4.126a}$$

$$D_x \tau_{n+1} \cdot \tau'_n = \frac{a}{b}(\tau_n \tau'_{n+1} - \tau_{n+1}\tau'_n). \tag{4.126b}$$

Then τ'_n may be shown to be the Nth-order wronskian

$$\tau'_n = [\widehat{\phi}(n), \widehat{\phi}(n+1), \ldots, \widehat{\phi}(n+N-1)], \tag{4.127}$$

where

$$\widehat{\phi}_j(n) = a\phi_j(n) + b\phi_j(n+1). \tag{4.128}$$

This solution τ'_n represents a solution with the same number of functions ϕ_j (the number of solitons) but with different phases.

Remarks

(1) The conditions $\mu = \lambda$, $\nu = \lambda^{-1}$, are necessary in order to allow the vacuum (zero-soliton) solutions $\tau_n = 1$, $\tau'_n =$ constant in (4.123).
(2) In the case of $a = 0, b = 1$, the solution τ'_n is the same as that obtained from the Bäcklund transformation for the KP equation. □

The verification of the solution (4.127) is made more difficult by the fact that $a \neq 0$. We first note that τ_n and τ_{n+1} have a number of expressions:

$$\begin{aligned}\tau_n &= [\phi(n), \phi(n+1), \ldots, \phi(n+N-1)]\\ &= a^{-N+1}[\widehat{\phi}(n), \widehat{\phi}(n+1), \ldots, \widehat{\phi}(n+N-2), \phi(n+N-1)]\end{aligned} \tag{4.129a}$$

$$= b^{-N+1}[\phi(n), \widehat{\phi}(n), \ldots, \widehat{\phi}(n+N-3), \widehat{\phi}(n+N-2)] \tag{4.129b}$$

$$= a^{-N}(\tau'_n - b[\widehat{\phi}(n), \widehat{\phi}(n+1), \ldots, \widehat{\phi}(n+N-2), \phi(n+N)]), \tag{4.129c}$$

$$\begin{aligned}\tau_{n+1} &= [\phi(n+1), \phi(n+2), \ldots, \phi(n+N)] \\ &= a^{-N+1}[\widehat{\phi}(n+1), \widehat{\phi}(n+2), \ldots, \widehat{\phi}(n+N-1), \phi(n+N)] \qquad (4.130a) \\ &= b^{-N+1}[\phi(n+1), \widehat{\phi}(n+1), \ldots, \widehat{\phi}(n+N-2), \widehat{\phi}(n+N-1)] \qquad (4.130b) \\ &= b^{-N}(\tau_n' - a[\phi(n), \widehat{\phi}(n+1), \ldots, \widehat{\phi}(n+N-2), \widehat{\phi}(n+N-1)]). \qquad (4.130c)\end{aligned}$$

Remark

This rewriting of the solution is easily carried out using elementary properties of determinants. For example, (4.129a) is derived in the following way:

$$\begin{aligned}\tau_n &= [\phi(n), \phi(n+1), \ldots, \phi(n+N-1)] \\ &= a^{-N+1}[a\phi(n), a\phi(n+1), \ldots, a\phi(n+N-2), \phi(n+N-1)] \\ &= a^{-N+1}[a\phi(n) + b\phi(n+1), \ldots, a\phi(n+N-2) \\ &\quad + b\phi(n+N-1), \phi(n+N-1)] \\ &= a^{-N+1}[\widehat{\phi}(n), \ldots, \widehat{\phi}(n+N-2), \phi(n+N-1)].\end{aligned}$$

□

Now, in order to prove (4.126a), we substitute the expression (4.129c) on the left-hand side and the expressions (4.129a), (4.130a) into the right-hand side. After computing derivatives, we obtain the Maya diagram expression for (4.126a):

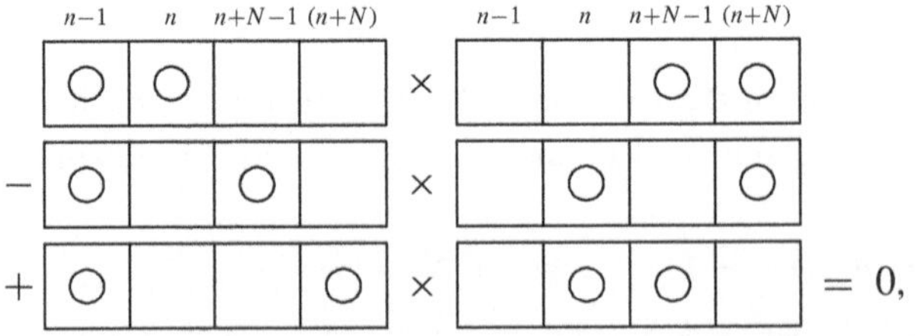

which is simply a Plücker relation. In the above Maya diagram expression, the labels m and (m) denote the presence of $\widehat{\phi}(m)$ and $\phi(m)$, respectively.

Following a similar procedure, we can prove (4.126b) by substituting (4.130c) into the left-hand side and (4.129b), (4.130b) into the right-hand side. Consequently, we obtain a Maya diagram expression equivalent to a Plücker relation.

Solution with one more soliton

For simplicity, we choose the free parameters to be $\lambda = 1,\ \mu = \nu = 0$. Then the Bäcklund transformation formulae are

$$D_s \tau_n \cdot \tau'_n = \tau_{n+1}\tau'_{n-1}, \tag{4.131a}$$

$$D_x \tau_{n+1} \cdot \tau'_n = -\tau_n \tau'_{n+1}. \tag{4.131b}$$

If τ_n is expressed in the wronskian form (4.124), then τ'_n is expressed as the $(N+1)$th-order wronskian

$$\tau'_n = (d_0, d_1, \ldots, d_{N-1}, d_N, N+1, N, \ldots, 2, 1). \tag{4.132}$$

This τ'_n represents a solution having one more of the functions ϕ_j (that is, one more soliton) than τ_n.

Remarks

(1) We chose the parameters to be $\lambda = 1$, $\mu = \nu = 0$, for the following reason. Substitution of a vacuum solution $\tau_n = 1$ and the one-soliton solution $\tau'_n = \phi(n)$ into the Bäcklund transformation formula gives the linear equations

$$\begin{aligned}\frac{\partial \phi_i(n)}{\partial s} &= -\lambda\phi_i(n-1) + \mu\phi_i(n),\\ \frac{\partial \phi_i(n)}{\partial x} &= \lambda^{-1}\phi_i(n+1) - \nu\phi_i(n).\end{aligned} \tag{4.133}$$

When $\mu = \nu = 0$, $\phi(n)$ gives a solution to the KP equation. Using scale transformations of the independent variables x, s, λ may be chosen to be unity.

(2) Since n represents the order of the x-derivative, τ_{n+1} and τ'_{n-1} may be denoted by

$$\tau_{n+1} = (d_1, d_2, \ldots, d_{N-1}, d_N, N, N-1, \ldots, 2, 1), \tag{4.134a}$$

$$\tau'_{n-1} = (d_{-1}, d_0, \ldots, d_{N-1}, N+1, N, \ldots, 2, 1). \tag{4.134b}$$

□

It is easy to prove (4.131a). Substituting (4.124), (4.132) into the left-hand side and (4.134a), (4.134b) into the right-hand side give the Maya diagram expression

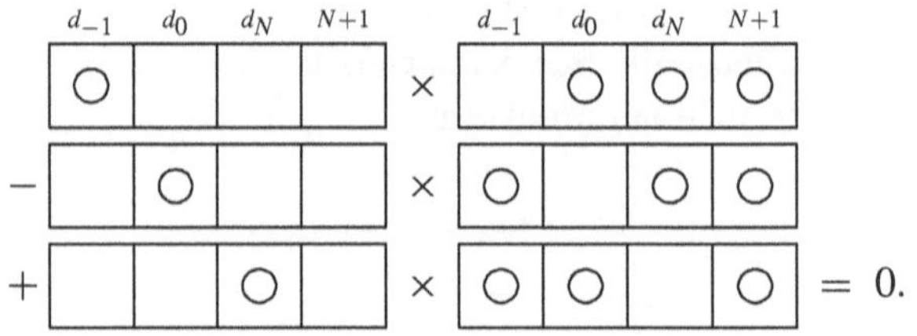

Equation (4.131b) can be proved in the same way.

4.7.2 Structure of the Bäcklund transformation for the Toda molecule equation

We noted in the previous section that the modified Toda equation is equivalent to the $2N$-wave interaction equations (4.116) if we adopt boundary conditions $u_0 = v_{N+1} = 0$, and take parameters $\mu = \nu = 0$. In the Bäcklund transformation formula,

$$D_s \tau_n \cdot \tau'_n = \lambda \tau_{n+1} \tau'_{n-1} - \mu \tau_n \tau'_n, \tag{4.135a}$$

$$D_x \tau_{n+1} \cdot \tau'_n = -\lambda^{-1} \tau_n \tau'_{n+1} + \nu \tau_{n+1} \tau'_n, \tag{4.135b}$$

the choice of parameters $\lambda = -b$, $\mu = 0$, $\nu = -a/b$, gives

$$D_s \tau_n \cdot \tau'_n = -b \tau_{n+1} \tau'_{n-1}, \tag{4.136a}$$

$$D_x \tau_{n+1} \cdot \tau'_n = b^{-1}(\tau_n \tau'_{n+1} - a \tau_{n+1} \tau'_n). \tag{4.136b}$$

The solutions τ_n in the above equations may be shown to be expressed in terms of bi-directional wronskians as

$$\tau_0 = 1, \ \tau_{-1} = 0, \ \tau_n = \det(\Psi_{i,j})_{0 \le i,j \le n-1}, \tag{4.137}$$

where

$$\Psi_{i,j} = \left(\frac{\partial}{\partial s}\right)^i \left(\frac{\partial}{\partial x}\right)^j \Psi(s,x), \quad \Psi(s,x) = \sum_{k=1}^{N+1} u_k(s) v_k(x), \tag{4.138}$$

and $u_k(s)$, $v_k(x)$ are arbitrary functions of s and x, respectively.

We may suppress reference to the s-derivatives in τ_n by writing

$$\tau_n \equiv [\Psi_0, \Psi_1, \ldots, \Psi_{n-1}] \tag{4.139}$$

for short, where Ψ_j denotes the column vector $(\Psi_{1,j}, \Psi_{2,j}, \ldots, \Psi_{n,j})^t$. Then τ'_n may be written as

$$\tau'_0 = 1, \quad \tau'_{-1} = 0,$$
$$\tau'_n = \det(\widehat{\Psi}_{i,j})_{0\le i,j\le n-1}, \tag{4.140}$$
$$\widehat{\Psi}_{i,j} = a\Psi_{i,j} + b\Psi_{i,j+1}, \tag{4.141}$$

and we can write

$$\tau'_n \equiv [\widehat{\Psi}_0, \widehat{\Psi}_1, \ldots, \widehat{\Psi}_{n-1}], \tag{4.142}$$

where

$$\widehat{\Psi}_j = a\Psi_j + b\Psi_{j+1}. \tag{4.143}$$

Remark

The necessity of the condition $\mu = 0$ is seen if one substitutes the vacuum solution $\tau_0 = \tau'_0 = 1$ into (4.135a). Also, we have chosen λ and ν so that (4.135b) coincides with (4.141) when $n = 0$. □

Since the subscript n indicates the size of the determinant, we rewrite τ_n and τ_{n+1} as

$$\begin{aligned}\tau_n &\equiv [\Psi_0, \Psi_1, \ldots, \Psi_{n-1}]\\ &= a^{-n+1}[\widehat{\Psi}_0, \widehat{\Psi}_1, \ldots, \widehat{\Psi}_{n-2}, \Psi_{n-1}] &(4.144\text{a})\\ &\equiv a^{-n}(\tau'_n - b[\widehat{\Psi}_0, \widehat{\Psi}_1, \ldots, \widehat{\Psi}_{n-2}, \Psi_n]), &(4.144\text{b})\end{aligned}$$

$$\begin{aligned}\tau_{n+1} &\equiv [\Psi_0, \Psi_1, \ldots, \Psi_{n-1}, \Psi_n]\\ &= a^{-n}[\widehat{\Psi}_0, \widehat{\Psi}_1, \ldots, \widehat{\Psi}_{n-2}, \widehat{\Psi}_{n-1}, \Psi_n]. &(4.145)\end{aligned}$$

In order to prove (4.136a), we introduce $(n+1)$th-, nth- and $(n-1)$th-order determinants D, $D\begin{bmatrix} i \\ j \end{bmatrix}$ and $D\begin{bmatrix} i & j \\ k & l \end{bmatrix}$:

$$D \equiv \begin{vmatrix} \widehat{\Psi}_{0,0} & \widehat{\Psi}_{0,1} & \cdots & \widehat{\Psi}_{0,n-1} & \Psi_{0,n} \\ \widehat{\Psi}_{1,0} & \widehat{\Psi}_{1,1} & \cdots & \widehat{\Psi}_{1,n-1} & \Psi_{1,n} \\ \vdots & \vdots & & \vdots & \vdots \\ \widehat{\Psi}_{n,0} & \widehat{\Psi}_{n,1} & \cdots & \widehat{\Psi}_{n,n-1} & \Psi_{n,n} \end{vmatrix}$$
$$\equiv \left[\widehat{\Psi}_0, \widehat{\Psi}_1, \ldots, \widehat{\Psi}_{n-2}, \widehat{\Psi}_{n-1}, \Psi_n\right] = a^n \tau_{n+1}, \tag{4.146a}$$

$$D\begin{bmatrix} i \\ j \end{bmatrix} = \text{determinant obtained by eliminating the } i\text{th row and the } j\text{th column of } D, \tag{4.146b}$$

$$D\begin{bmatrix} i & j \\ k & l \end{bmatrix} = \text{determinant obtained by eliminating the } i\text{th and } j\text{th rows and the } k\text{th and } l\text{th columns of } D. \tag{4.146c}$$

By using the above notation, τ_n' and its derivatives may be written as

$$\tau_n' = D\begin{bmatrix} n+1 \\ n+1 \end{bmatrix}, \tag{4.147a}$$

$$\tau_{n-1}' = D\begin{bmatrix} n & n+1 \\ n & n+1 \end{bmatrix}, \tag{4.147b}$$

$$\frac{\partial \tau_n'}{\partial s} = D\begin{bmatrix} n \\ n+1 \end{bmatrix}, \tag{4.147c}$$

$$\begin{aligned} D_s\tau_n \cdot \tau_n' &= D_s a^{-n}\left(\tau_n' - b\left[\widehat{\Psi}_0, \widehat{\Psi}_1, \ldots, \widehat{\Psi}_{n-2}, \Psi_n\right]\right) \cdot \tau_n' \\ &= -a^{-n}b\left\{D\begin{bmatrix} n \\ n \end{bmatrix} D\begin{bmatrix} n+1 \\ n+1 \end{bmatrix} - D\begin{bmatrix} n+1 \\ n \end{bmatrix} D\begin{bmatrix} n \\ n+1 \end{bmatrix}\right\}. \end{aligned} \tag{4.147d}$$

Therefore, (4.136a) is simply the Jacobi identity

$$D\begin{bmatrix} n \\ n \end{bmatrix} D\begin{bmatrix} n+1 \\ n+1 \end{bmatrix} - D\begin{bmatrix} n+1 \\ n \end{bmatrix} D\begin{bmatrix} n \\ n+1 \end{bmatrix} = D\begin{bmatrix} n & n+1 \\ n & n+1 \end{bmatrix} D. \tag{4.148}$$

The rest of this section is devoted to a proof of (4.136b). We first rewrite this as

$$\left[a\tau_{n+1} + b\left(\frac{\partial \tau_{n+1}}{\partial x}\right)\right]\tau_n' - b\tau_{n+1}\left(\frac{\partial \tau_n'}{\partial x}\right) - \tau_{n+1}'\tau_n = 0, \tag{4.149}$$

and use the expressions

$$\begin{aligned} a\tau_{n+1} + b\left(\frac{\partial \tau_{n+1}}{\partial x}\right) &= [\Psi_0, \Psi_1, \ldots, \Psi_{n-1}, \widehat{\Psi}_n] \\ &= a^{-n+1}[\widehat{\Psi}_0, \widehat{\Psi}_1, \ldots, \widehat{\Psi}_{n-2}, \Psi_{n-1}, \widehat{\Psi}_n], \end{aligned} \tag{4.150a}$$

$$b\tau_{n+1} = a^{-n+1}[\widehat{\Psi}_0, \widehat{\Psi}_1, \ldots, \widehat{\Psi}_{n-2}, \Psi_{n-1}, \widehat{\Psi}_{n-1}]. \tag{4.150b}$$

Substitution of these expressions and (4.144a) into (4.149) yields

$$\begin{aligned}
&[\widehat{\Psi}_0, \widehat{\Psi}_1, \ldots, \widehat{\Psi}_{n-2}, \Psi_{n-1}, \widehat{\Psi}_n][\widehat{\Psi}_0, \widehat{\Psi}_1, \ldots, \widehat{\Psi}_{n-2}, \Psi_{n-1}] \\
&\quad - [\widehat{\Psi}_0, \widehat{\Psi}_1, \ldots, \widehat{\Psi}_{n-2}, \Psi_{n-1}, \widehat{\Psi}_{n-1}][\widehat{\Psi}_0, \widehat{\Psi}_1, \ldots, \widehat{\Psi}_{n-2}, \widehat{\Psi}_n] \\
&\quad - [\widehat{\Psi}_0, \widehat{\Psi}_1, \ldots, \widehat{\Psi}_{n-2}, \widehat{\Psi}_{n-1}, \widehat{\Psi}_n][\widehat{\Psi}_0, \widehat{\Psi}_1, \ldots, \widehat{\Psi}_{n-2}, \Psi_{n-1}] = 0.
\end{aligned} \tag{4.151}$$

The above determinants can be regarded as wronskians with respect to x-derivatives. Hence, by introducing the pfaffian entry $(b_n, m) = (\partial^n/\partial x^n)\widehat{\Psi}_m$, (4.151) results in the Maya diagram expression

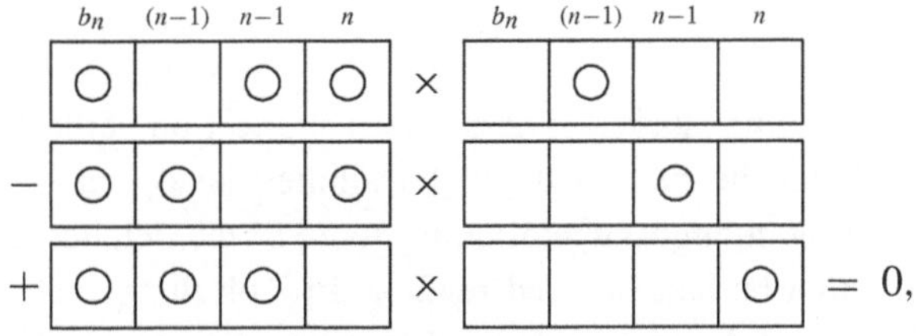

which is simply a pfaffian identity.

Snake's legs

Viewing the above proof, some readers might suppose that the techniques used are so complicated that they could not possibly discover this result. However, it is easier than they might imagine. The author first checked by hand the expressions he had guessed in the case $n = 0, 1$ and then investigated the cases $n = 2, 3, \ldots$ using computer algebra. After confirming that the guess is likely to be correct, he finally considered the proof of the general case. Confirmation at each step of the proof using computer algebra also reduces the possibility of errors. □

We have discussed Bäcklund transformations and the structure of the solutions of the bilinear KP, BKP and Toda equations, which are considered as the most fundamental soliton equations, and we have shown that the Bäcklund transformation formulae are, in each case, simply pfaffian identities.

Afterword

No sooner had the author started to describe an application of the direct method that he realized that he had used up the allotted space on the fundamentals. Even though he thought of shortening some of the detailed explanations, he remembered that he had suffered reading difficult mathematics books because of their terse style, and so decided to retain the seemingly superfluous remarks.

Let us briefly mention some topics not discussed in the book.

(1) Fundamental soliton equations such as the KP, BKP and Toda equations and their Bäcklund transformation formulae may be regarded as 'atoms' for constructing various kinds of soliton equations. Combination of these equations generate many other soliton equations and their solutions. Modern science has been able to understand the properties of materials by decomposing them into their constituents, or atoms, and has managed to create new materials by combining different atoms. It is a pity that lack of space prevented the author explaining how to construct new soliton equations from the above atoms. For example, the KP equation and its Bäcklund transformation formula may also be considered as the bilinear form of the nonlinear Schrödinger equation. In this way, we can construct the dromion solution (two-dimensionally localized soliton) for the Davey–Stewartson equation. It should also be noted that apparently different nonlinear partial differential equations are frequently transformed into the same bilinear form.

(2) Since Bäcklund transformation formulae are also considered as soliton equations (the first modified equations), their Bäcklund transformation formulae (second modified equations) also exist. Furthermore, Bäcklund transformation formulae for the second modified equation, that is, third

modified equations, can also be constructed and they make an infinite hierarchy. We may expect that there may be interesting soliton equations among this hierarchy.

(3) What kind of soliton equations are generated from Bäcklund transformation formulae of the coupled KP equation?

(4) The word 'pfaffian' first appeared in connection with solitons in ref. [63].

(5) The author believes that all soliton equations can be discretized. Therefore, they generate an infinite number of nonlinear partial difference equations equipped with an infinite number of conserved quantities and exact solutions. This belief is supported by the remarkable fact that τ-functions of the KP equation and the difference KP equation coincide with each other.

The interaction of solitons illustrated in Figure 1.7 was calculated using the nonlinear partial difference equation (difference scheme)

$$u_n^{t+1} - u_n^t = \delta \left\{ u_{n-1}^{t+1} u_n^t - u_n^{t+1} u_{n+1}^t \right\},$$

where n, t and δ stand for the lattice position, the time and the length of the time interval, respectively. This equation has an infinite number of conserved quantities under periodic boundary conditions. However, it was recently discovered that the time evolution of this integrable difference equation exhibits chaotic behaviour under the periodic boundary condition $u_{n-1}^t = \alpha$, $u_n^t = x^t$, $u_{n+1}^t = y^t$, $u_{n+2}^t = \beta$.

(6) Shin'ichi Oishi [64] studied how to solve initial value problems for partial differential equations by using their bilinear forms. According to this author, if a solution may be expressed as a determinant, the initial value problem can be solved using the bilinear form.

(7) It should also be noted that the late Nobuo Yajima applied bilinear methods to a perturbed system [65].

(8) There is a strong relationship between soliton equations and the special functions used in physics. In a sense, soliton theory can be considered as a theory of special functions. Akira Nakamura has been investigating special functions from this point of view. Ref. [66] reviews Nakamura's results obtained up to 1988, and later results appear in the following papers: [67] (Airy function, Hermite polynomials); [68] (associated Legendre functions); [69] (associated Legendre functions, generalized Laguerre functions); [70] (Jacobi polynomials); [71] (Gauss hyper-geometric functions); [72] (Legendre polynomials). The full implications of these results are still unknown.

(9) Finally, two other explanations of the bilinear method are mentioned [73,74]. The main feature of the book by Matsuno [73] is a detailed explanation of a bilinearization of the Benjamin–Ono equation. Also, the survey article by Nimmo [74] explains how vertex operators generate soliton solutions and describes the group theoretical significance of bilinear equations.

References

[1] G. A. Baker and J. L. Gammel (eds.). *The Padé Approximant in Theoretical Physics* (New York: Academic Press, 1970).
[2] N. J. Zabusky and M. D. Kruskal. *Phys. Rev. Lett.* **15** (1965) 240.
[3] C. S. Gardner, J. M. Greene, M. D. Kruskal and R. M. Miura. *Phys. Rev. Lett.* **19** (1967) 1095.
[4] E. Hopf. *Commun. Pure. Appl. Math.* **3** (1950) 201.
[5] G. L. Lamb, Jr. 'Bäcklund transformations at the turn of the century', in *Bäcklund Transformations*, ed. R. M. Miura. Lecture Notes in Mathematics, vol. 515 (Berlin: Springer, 1976).
[6] A. Hasegawa. *Phys. Lett.* **47A** (1974) 165.
[7] H. Hashimoto. *Proc. Jpn Acad.* **50** (1974) 623.
[8] R. M. Miura. 'Introduction', in *Bäcklund Transformations*, ed. R. M. Miura. Lecture Notes in Mathematics, vol. 515 (Berlin: Springer, 1976).
[9] R. Hirota. *Progr. Theor. Phys.* **52** (1974) 1498.
[10] R. Hirota. *Phys. Rev. Lett.* **27** (1971) 1192.
[11] M. Kashiwara, M. Jimbo, E. Date and T. Miwa. *Sugaku*, **34** (1982) no. 1. (In Japanese.)
[12] M. Sato. *RIMS Kokyuroku* **439** (1981) 30.
[13] Y. Ohta, J. Satsuma, D. Takahashi and T. Tokihiro. *Progr. Theor. Phys. Suppl.* **94** (1988) 210.
[14] E. Date, M. Kashiwara, M. Jimbo and T. Miwa. *Non-linear Integrable Systems – Classical Theory and Quantum Theory*, eds. M. Jimbo and T. Miwa (Singapore: World Scientific, 1983), p. 39.
[15] M. Jimbo and T. Miwa. *Publ. RIMS, Kyoto Univ.* **19** (1983) 39.
[16] R. Hirota and M. Itoh. *Introduction to REDUCE* (Tokyo: Science, 1989). (In Japanese.)
[17] R. Hirota 'Fundamental properties of the binary operators in soliton theory and their generalization', in *Dynamical Problems in Soliton Systems*, ed. S. Takeno Springer Series in Synergetics, vol. 30 (Berlin: Springer, 1985).
[18] K. Nozaki and N. Bekki. *J. Phys. Soc. Jpn* **53** (1984) 1581.
[19] K. Sawada and T. Kotera. *Progr. Theor. Phys.* **51** (1974) 1355.
[20] R. Hirota and J. Satsuma. *Progr. Theor. Phys. Suppl.* **59** (1976) 64.

[21] R. Hirota. 'Direct method of finding exact solutions of nonlinear evolution equations', in *Bäcklund Transformations*, ed. R. M. Miura. Lecture Notes in Mathematics, vol. 515 (Berlin: Springer, 1976).
[22] R. Hirota. 'Direct methods in soliton theory', in *Solitons*, eds. R. K. Bullough and P. J. Caudrey. Topics in Current Physics, vol. 17 (Berlin: Springer-Verlag, 1980).
[23] R. Hirota. *J. Phys. Soc. Jpn* **33** (1972) 1456.
[24] R. Hirota. *J. Phys. Soc. Jpn* **33** (1972) 1459.
[25] A. Nakamura. *J. Phys. Soc. Jpn* **47** (1972) 1701.
[26] J. Hietarinta. *J. Math. Phys.* **28** (1987) 1732.
[27] J. Hietarinta. *J. Math. Phys.* **28** (1987) 2094.
[28] J. Hietarinta. *J. Math. Phys.* **28** (1987) 2586.
[29] J. Hietarinta. *J. Math. Phys.* **29** (1988) 628.
[30] E. R. Caieniello. *Combinatorics and Renormalization in Quantum Field Theory* (New York: Benjamin, 1973).
[31] R. Hirota. *J. Phys. Soc. Jpn* **58** (1989) 2285.
[32] M. Sato and M. Noumi. *Sugaku Kokyuroku* (Tokyo: Sophia University, 1984), no. 18. (In Japanese.)
[33] T. Takagi. *Lecture in Algebra* (Tokyo: Kyoritsu, 1965), p. 254. (In Japanese.)
[34] M. Toda and K. Asano. *Matrices and Linear Transformations* (Tokyo: Iwanami, 1989).
[35] Y. Ohta. *Bilinear Theory of Soliton*, Ph.D. Thesis (Faculty of Engineering, University of Tokyo, 1992). (In Japanese.)
[36] A. Nakamura. *J. Phys. Soc. Jpn* **58** (1989) 412.
[37] J. Satsuma. *J. Phys. Soc. Jpn* **46** (1979) 359.
[38] N. C. Freeman and J. J. C. Nimmo. *Phys. Lett.* **95A** (1983) 1.
[39] J. J. C. Nimmo and N. C. Freeman. *Phys. Lett.* **95A** (1983) 4.
[40] B. B. Kadomtsev and V. I. Petviashvili. *Sov. Phys. Doklady* **15** (1970) 539.
[41] J. Satsuma. *J. Phys. Soc. Jpn* **48** (1976) 286.
[42] S. Oishi. *J. Phys. Soc. Jpn* **48** (1980) 639.
[43] R. Hirota. *J. Phys. Soc. Jpn* **55** (1986) 2137.
[44] A. Nakamura. *J. Phys. Soc. Jpn* **58** (1989) 412.
[45] M. J. Ablowitz and H. Segur. *Solitons and the Inverse Scattering Transform* (Philadelphia: SIAM, 1981).
[46] R. Hirota. *J. Phys. Soc. Jpn* **58** (1989) 2705.
[47] R. Hirota and J. Satsuma. *Phys. Lett.* **85A** (1981) 407.
[48] R. Hirota and Y. Ohta. *J. Phys. Soc. Jpn* **60** (1990) 798.
[49] R. Hirota, Y. Ohta and J. Satsuma. *J. Phys. Soc. Jpn* **57** (1988) 1901.
[50] A. Nakamura. *J. Phys. Soc. Jpn* **57** (1988) 3309.
[51] R. Hirota. *J. Phys. Soc. Jpn* **56** (1987) 4285.
[52] A. N. Leznov and M. V. Saveliev. *Physica* **3D1&2** (1981) 62.
[53] R. Hirota. *J. Phys. Soc. Jpn* **57** (1988) 436.
[54] R. Hirota, Y. Ohta and J. Satsuma. *Progr. Theor. Phys. Suppl.* **94** (1988) 59.
[55] G. Darboux. *Théorie générale des surfaces, II* (New York: Chelsea Publishing Company, 1972).
[56] Y. Kametaka. *Japan J. Appl. Math.* **2** (1985) 241.
[57] A. C. Scott, F. Y. F. Chu and W. McLaughlin. *Proc. IEEE* **61** (1973) 1443.

[58] C. Rogers and W. E. Shadwick. *Bäcklund Transformations and Their Applications* (New York: Academic Press, 1982).
[59] R. Hirota. *Progr. Theor. Phys. Suppl.* **52** (1974) 1948.
[60] M. Wadati, H. Sanuki and K. Konno. *Progr. Theor. Phys.* **53** (1975) 418.
[61] R. M. Miura. *J. Math. Phys.* **9** (1968) 1202.
[62] R. Hirota, Y. Ohta and J. Satsuma. *Progr. Theor. Phys. Suppl.* **94** (1988) 59.
[63] E. Date, M. Jimbo, M. Kashiwara and T. Miwa. *Physica* **4D** (1982) 343.
[64] S. Oishi. *J. Phys. Soc. Jpn* **48** (1980) 639.
[65] N. Yajima. *J. Phys. Soc. Jpn* **51** (1982) 1298.
[66] A. Nakamura. *Prog. Theor. Phys. Suppl.* **94** (1988) 195.
[67] A. Nakamura. *J. Phys. Soc. Jpn* **58** (1989) 412.
[68] A. Nakamura. *J. Phys. Soc. Jpn* **58** (1989) 2687.
[69] A. Nakamura. *J. Phys. Soc. Jpn* **59** (1990) 1553.
[70] A. Nakamura. *J. Phys. Soc. Jpn* **59** (1990) 3101.
[71] A. Nakamura. *J. Phys. Soc. Jpn* **59** (1990) 4272.
[72] A. Nakamura and Y. Ohta. *J. Phys. Soc. Jpn* **60** (1991) 1835.
[73] M. Matsuno. *Bilinear Transformation Method* (New York: Academic Press, 1984).
[74] J. J. C. Nimmo. 'Hirota's method', in *Soliton Theory: a survey of results*, ed. A. P. Fordy. (Manchester: Manchester University Press, 1990).

Index

In this index *italics* is used to specify page numbers for major references such as definitions or sections on the topic.